Introduction to Matrices and Vectors

Introduction to Matrices and Vectors

Jacob T. Schwartz
Professor of Mathematics
Institute of Mathematical Sciences
New York University

DOVER PUBLICATIONS, INC.
Mineola, New York

Copyright

Published in Canada by General Publishing Company, Ltd., 30 Lesmill Road, Don Mills, Toronto, Ontario.

Bibliographical Note

This Dover edition, first published in 2001, is an unabridged and unaltered republication of the work originally published by the McGraw-Hill Book Company, Inc., in 1961.

Library of Congress Cataloging-in-Publication Data

Schwartz, Jacob T.
Introduction to matrices and vectors / Jacob T. Schwartz.—Dover ed.
p. cm.
Originally published: New York : McGraw-Hill Book Co., 1961.
Includes index.
ISBN 0-486-42000-0 (pbk.)
1. Matrices. 2. Vector analysis. I. Title.

QA188 .S247 2001
512.9'434—dc21

2001028840

Manufactured in the United States of America
Dover Publications, Inc., 31 East 2nd Street, Mineola, N.Y. 11501

PREFACE

The theory of matrices occupies a strategic position both in pure and in applied mathematics. The most famous application is doubtless the application to atomic physics; the prevalence of such terms as *matrix mechanics, scattering matrix, spin matrices, annihilation and creation matrices* gives vivid testimony to the decisive importance of matrix theory for the modern physicist. In economics we have the *input-output matrix*, computed at great expense by the Bureau of Labor Statistics, as well as the game-theoretical *payoff matrix.* In statistics we have *transition matrices*, as well as many other sorts of matrices; in engineering, the *stress matrix*, the *strain matrix*, and many other important matrices. Many of the numerical problems attacked with the aid of modern high-speed digital computing machines are expressed in terms of matrices; *matrix inversion, eigenvalues of matrices*, and so forth, are everyday terms at any major computation installation. In pure mathematics, also, matrices play a great role. *Matrix representation* is a familiar problem for the algebraist, who has by now proved that almost every abstract algebraic system can be given concrete representation by use of matrices. The *Jacobian matrix* is a familiar object to the specialist in analysis, who is also accustomed to write systems of equations as matrix equations. And so forth. The subject surely needs no justification.

But even if the subject requires no justification, the particular treatment advanced by a given author surely does. In writing the present book, I have emphasized the concrete and computational, convinced that students at a more elementary level are better able to comprehend general theories when so presented than to fathom the significance of possibly very short, but still highly abstract, proofs and principles. From the beginning then, matrices are specific objects with which the student is to compute; exercises attached to each short section give practice in this. The general theory develops gradually out of specific computations; as it develops, it is made clear that computation with matrices has very much the same structure as ordinary numerical computation. This should make it possible for the good senior high school student, or the average college freshman and sophomore, to pass smoothly from the familiar computationally flavored algebra of high school to the structure-oriented viewpoint of modern algebra. The third chapter, with its emphasis on the "minimal equation," already has much of the flavor of abstract "commutative algebra"—but the computational thread is preserved through a section on the computation of the minimal equation.

Chapters 4 to 8 give various applications to more specific situations of the general algebraic procedures developed in the first three chapters.

The present book can be used for a variety of courses. There has, of late, been a tendency to devote part of the twelfth year of high school mathematics to the study of some "algebraic" topic. Because of its computational flavor and its importance in applications, there can be no doubt that matrix theory is to be preferred to any other algebraic topic at the high school level. A one-semester high school course might consist of Chapters 1, 2, 4, 5, and either 6, 3, or, time and the class permitting, both.

A normal college freshman or sophomore one-semester course might cover Chapters 1 to 7.

Chapter 8 goes beyond the purely algebraic and introduces the "infinitesimal" note lately developed so successfully in the theory of "Banach algebras."

Thanks are due to many persons. My colleagues at the SMSG Conference at Boulder, Colorado, in the summer of 1959, where the idea for the present book was born, were full of suggestions. Particular thanks are due to Esther Gasset of the Claremore High School, Claremore, Oklahoma, from whose keen pedagogical sense, great experience, and down-to-earth soundness I profited greatly. Thanks finally are due to Prof. E. G. Begle of Yale University, but for whom the present book would never have appeared.

Jacob T. Schwartz

CONTENTS

Chapter 1

DEFINITION, EQUALITY, AND ADDITION OF MATRICES

1-1. Introduction

As we have done more and more sophisticated mathematics in our previous studies, we have had occasion to use more and more sophisticated kinds of "numbers." We began with *positive whole numbers*, like 1, 2, 3, ... Then, in order to make subtractions like $3 - 7$ possible, the *negative whole numbers*, like -1, -2, -3, . . . , had to be introduced. Next, in order to make it possible to divide any two numbers, *fractions* were invented: $\frac{1}{2}$, $-\frac{2}{3}$, $-\frac{157}{321}$, etc. This still did not bring us to the end of our story; for, in order that every number have a square root, a cube root, a logarithm, a sine, and so forth, it was necessary to invent still more numbers: the *infinite decimals*, or *real numbers*, like 1.4142 . . . , 3.1415928 . . . , 0.13131313 . . . . Finally, in order that even negative numbers have square roots, it was necessary to invent *complex numbers*, like $3 + 2i$, $1 + \pi i$, $-\frac{1}{2} + \frac{1}{37}i$.

Whenever there seemed to be good reason to do so, we have *invented* new kinds of numbers. For instance, in inventing complex numbers, we began not with the numbers but with a *purpose:*

to find a system of numbers, including all the real numbers, in which every number had a square root. Once we have made such an invention, it should not be hard for us to realize that there is no reason to stop inventing; that is, there is no reason why other kinds of numbers should not be useful for other purposes. There is *no reason why we should not hope to invent many different kinds of new numbers.*

Of course, as an inventor, it is easy to invent things that do not work, but harder to invent things that do work; easy to invent things that are useless, but hard to invent things that are useful. The same is true about the invention of new kinds of numbers. The hard thing is to invent useful kinds of numbers, and kinds of numbers that work. It is easier to make inventions than to make successful inventions. Nevertheless, a large variety of more or less successful new kinds of numbers have been invented by mathematicians. In this course you are going to study one of the most successful of these new kinds of numbers: the *matrices.*

Before you are told what matrices are, it is well to emphasize their importance. Matrices are useful in almost every branch of science and engineering. A great number of the computations made on the giant "electronic brains" are computations with matrices. Many problems in statistics are expressed in terms of matrices. Matrices come up in the mathematical problems of economics. Matrices are extremely important in the study of atomic physics; indeed, atomic physicists express almost all their problems in terms of matrices. It would not be an exaggeration to say that the algebra of matrices is the language of atomic physics. Many other kinds of algebra, like complex-number algebra (and like vector algebra, which some of you may already have studied), can be explained very easily in terms of matrices. So, in studying matrices, you will be studying one of the most useful and important and also one of the most interesting branches of mathematics.

1-2. What Matrices Are

A matrix is, basically, a very simple sort of thing.

A *matrix* is a square array of *real* numbers, arranged in rows and columns. If there are n numbers in each row and column of the array, the matrix is said to be an n *by* n matrix, or a matrix of *size* n.

"Array" means the same thing as "arrangement." Thus a matrix is an arrangement of numbers in a square. The *matrix* is the *arrangement* of many numbers, and not any single number.

For example, the arrays

$$\begin{bmatrix} 2 & 3 \\ 1 & -2 \end{bmatrix}$$

and

$$\begin{bmatrix} 2 & 1 \\ 3 & 4 \end{bmatrix}$$

are 2×2 matrices, or matrices of size 2. The array

$$\begin{bmatrix} -1 & 2 & 3 \\ 4 & -5 & 6 \\ 7 & 8 & -9 \end{bmatrix}$$

is a 3×3 matrix, or a matrix of size 3. Another 2×2 matrix is

$$\begin{bmatrix} 3.14 & 0 \\ \frac{10}{11} & 42 \end{bmatrix}.$$

Another 3×3 matrix is

$$\begin{bmatrix} 1 & 0 & 0 \\ 99 & 1 & 0 \\ 1.42 & 99 & 1 \end{bmatrix}.$$

Our idea is to consider such an array of many numbers as a *single object*, an *array*, a *matrix* and to give the whole array a single name or symbol. Thus, we might call the first of the above arrays A, the second B, the third C, etc. This procedure might at first seem pointless. As we shall realize more and more clearly in the course of our work, it is not pointless at all. It has the following very important consequence: by regarding a square array of numbers as constituting a single object, a matrix, *we will be able to handle large sets of numbers as single units, thereby simplifying the statement of complicated relationships.*

A matrix is a square array of numbers. The individual numbers which occur in this array are called the *entries* of the matrix. We shall identify particular entries in a matrix by specifying the horizontal *row* and vertical *column* to which they belong.

DEFINITION

If A is a matrix, the symbol $[A]_{i,j}$ will denote the entry in the ith row and jth column of the matrix A.

Thus, for instance, if M is the 5×5 matrix

$$\begin{bmatrix} 1 & 2 & 3 & 4 & 5 \\ 8 & 10 & 12 & 14 & 16 \\ -1 & -3 & -5 & 6 & 3 \\ 0 & 3 & -7 & 8 & 7 \\ 4 & 2 & 1 & -9 & 0 \end{bmatrix},$$

the entry in the third row, second column is -3. Thus $[M]_{3,2} = -3$. The entry in the second row, third column is 12. Thus $[\mathrm{M}]_{2,3} = 12$. All the entries in M are whole numbers (either positive or negative). Just two of the entries are zero.

EXERCISES

1. Let I be the matrix

$$\begin{bmatrix} 1 & 0 & 0 & 0 & 0 \\ 0 & 1 & 0 & 0 & 0 \\ 0 & 0 & 1 & 0 & 0 \\ 0 & 0 & 0 & 1 & 0 \\ 0 & 0 & 0 & 0 & 1 \end{bmatrix}.$$

(*a*) What is $[I]_{1,2}$? What is $[I]_{1,1}$?

(*b*) When is $[I]_{i,j} \neq 0$? When is $[I]_{i,j} = 0$?

(*c*) Name the entries in the third row.

(*d*) Name the entries in the second column.

2. Write a 3×3 matrix all of whose entries are whole numbers. Write a 4×4 matrix none of whose entries are whole numbers. Write a 5×5 matrix having all positive entries in its first two rows and all negative entries in its last three rows.

3. How many entries are there in a 2×2 matrix? In a 3×3 matrix? In an $n \times n$ matrix?

1-3. Equality of Matrices. Specification of Matrices. The Zero Matrix

DEFINITION

Two matrices A and B are said to be equal if

(*a*) A and B are of the same size.

(*b*) All the entries of A are the same as the corresponding entries of B.

Thus A and B can never be equal if A is a 9×9 matrix and B is a 10×10 matrix, since they are of different sizes. If A and B are of the same size n, then by the above definition, they are equal if and only if their corresponding entries are equal, that is, if and only if

$$[A]_{i,j} = [B]_{i,j} \qquad \text{for every } i \text{ and } j \text{ between 1 and } n.$$

Using the foregoing definition of equality, we can express certain relationships more compactly. For example, the equation

$$\begin{bmatrix} x+y & a+b \\ x-y & a-b \end{bmatrix} = \begin{bmatrix} 5 & -1 \\ 1 & 3 \end{bmatrix}$$

can be written in place of the four equations

$$\begin{aligned} x + y &= 5, \\ x - y &= 1, \\ a + b &= -1, \\ a - b &= 3. \end{aligned}$$

The nine equations

$$x + x = 2, \quad x + y = 3, \quad x + z = 4, \quad y + x = 3,$$
$$y + y = 4, \quad y + z = 5, \quad z + x = 4, \quad z + y = 5, \quad z + z = 6$$

can be expressed in matrix form as

$$\begin{bmatrix} x + x & x + y & x + z \\ y + x & y + y & y + z \\ z + x & z + y & z + z \end{bmatrix} = \begin{bmatrix} 2 & 3 & 4 \\ 3 & 4 & 5 \\ 4 & 5 & 6 \end{bmatrix}.$$

It is clear that to know a matrix is exactly the same thing as to know all its entries. These entries can be any real numbers we like. Thus, a matrix is determined when we specify all its entries. To determine a 2×2 matrix, we must specify 4 entries: to specify a 3×3 matrix, we must give 9 entries. To specify an $n \times n$ matrix, we must specify n^2 entries. These entries may be specified in any way we like.

Thus, we may determine a 2×2 matrix A by specifying that $A_{1,1} = 0$, $A_{1,2} = \pi$, $A_{2,1} = \sqrt{2}$, $A_{2,2} = \sqrt[3]{3}$. The matrix A is then

$$A = \begin{bmatrix} 0 & \pi \\ \sqrt{2} & \sqrt[3]{3} \end{bmatrix}.$$

We may determine a 4×4 matrix B by specifying that $[B]_{i,j} = 0$ if $i < j$ and $[B]_{i,j} = 1$ if $i \geq j$. The matrix B is then

$$B = \begin{bmatrix} 1 & 0 & 0 & 0 & 0 \\ 1 & 1 & 0 & 0 & 0 \\ 1 & 1 & 1 & 0 & 0 \\ 1 & 1 & 1 & 1 & 0 \\ 1 & 1 & 1 & 1 & 1 \end{bmatrix}.$$

A very simple, but important, $n \times n$ matrix is the $n \times n$ *zero matrix*. This is simply the $n \times n$ matrix all of whose entries are zero. Thus, the 2×2 zero matrix is

$$\begin{bmatrix} 0 & 0 \\ 0 & 0 \end{bmatrix},$$

the 3×3 zero matrix is

$$\begin{bmatrix} 0 & 0 & 0 \\ 0 & 0 & 0 \\ 0 & 0 & 0 \end{bmatrix},$$

etc. The $n \times n$ zero matrix can be denoted by the symbol 0_n. Thus

$$0_2 = \begin{bmatrix} 0 & 0 \\ 0 & 0 \end{bmatrix},$$
$$0_3 = \begin{bmatrix} 0 & 0 & 0 \\ 0 & 0 & 0 \\ 0 & 0 & 0 \end{bmatrix},$$

etc. But often it will be clear from the context that we are talking about matrices of some definite size. If this is the case, we may leave off the subscript of the zero matrix and write it simply as 0. This will be done only when the reader knows what size matrix is meant, or when it does not matter what size matrix is meant. Hence, it should never lead to confusion.

EXERCISES

1. Specify a 5×5 matrix all of whose entries are positive.

2. Specify 100 different matrices all of whose entries are 1.

3. Specify a nonzero 3×3 matrix whose entries satisfy $[A]_{i,j} = [A]_{j,i}$. Specify a nonzero 3×3 matrix whose entries satisfy $[A]_{i,j} = -[A]_{j,i}$. If A is such a matrix, what must $[A]_{1,1}$ be? What must $[A]_{2,2}$ and $[A]_{3,3}$ be?

1-4. Addition of Matrices

By now we have defined matrices and studied some of their elementary properties. But we have not really "made them work." To do this, we must give rules for adding and multiplying matrices. To see that this is the case, consider what was done with *complex numbers* in the earlier grades. Complex numbers were initially defined just as ordered pairs (x,y) of real numbers. If one stopped there, one certainly could not claim that complex numbers were very interesting. What makes complex numbers useful and interesting is that we are able to define addition and multiplication of complex numbers in a suitable way. Once this is done, we deal not merely with individual complex numbers, but with laws for their addition and multiplication, from which we can arrive at laws of exponents, study polynomials and equations, etc. In short, we have not merely individual complex numbers, which would be rather trivial, but a whole *algebra* of complex numbers, which is both useful and interesting.

The same remark applies to matrices. To give the study of matrices any real content, we must define "sum" and "product" for matrices.

In this section, we define and study sums of matrices. Products of matrices will be defined and studied later.

DEFINITION

By the *sum* of two $n \times n$ matrices A and B, we simply mean the matrix C whose entry in the ith row and jth column is the sum of $[A]_{i,j}$ and $[B]_{i,j}$.

A shorter way of giving the same definition is: the sum matrix $A + B$ of two matrices A and B is defined by the formula

$$[A + B]_{i,j} = [A]_{i,j} + [B]_{i,j}. \tag{1}$$

To write the definition in this very explicit way will help us greatly in what follows, since it makes the defining property of a sum matrix dramatically visible and shows us that our notation for the entries of a matrix is well adapted to the algebraic purposes for which it is to be used. The student should study this definition most carefully. He should also be very careful to make sure that he understands exactly how and why formula (1) expresses precisely the same thought as the more verbal definition which precedes it.

Thus, matrices of the same size are to be added simply by adding each element of one matrix to the corresponding element of the other matrix. Thus:

$$\begin{bmatrix} 1 & 2 \\ 3 & 4 \end{bmatrix} + \begin{bmatrix} 5 & 6 \\ 7 & 8 \end{bmatrix} = \begin{bmatrix} 1+5 & 2+6 \\ 3+7 & 4+8 \end{bmatrix} = \begin{bmatrix} 6 & 8 \\ 10 & 12 \end{bmatrix}$$

$$\begin{bmatrix} 1+\sqrt{2} & 0 \\ -1 & 1-\sqrt{2} \end{bmatrix} + \begin{bmatrix} -\sqrt{2} & 0 \\ 1 & +\sqrt{2} \end{bmatrix} = \begin{bmatrix} 1 & 0 \\ 0 & 1 \end{bmatrix}$$

$$\begin{bmatrix} 1 & 2 & 3 \\ 4 & 5 & 6 \\ 7 & 8 & 9 \end{bmatrix} + \begin{bmatrix} 0.1 & 0.2 & 0.3 \\ 0.4 & 0.5 & 0.6 \\ 0.7 & 0.8 & 0.9 \end{bmatrix} = \begin{bmatrix} 1.1 & 2.2 & 3.3 \\ 4.4 & 5.5 & 6.6 \\ 7.7 & 8.8 & 9.9 \end{bmatrix}$$

etc.

We shall add two matrices *only* if they are of the same size. Two matrices will never be added if they are of different sizes. We will not even define a rule by which matrices of different sizes could be added.

EXERCISES

1. What is

$$\begin{bmatrix} \frac{1}{2} & \frac{1}{3} \\ \frac{1}{4} & \frac{1}{5} \end{bmatrix} + \begin{bmatrix} \frac{1}{6} & \frac{1}{7} \\ \frac{1}{8} & \frac{1}{9} \end{bmatrix}?$$

2. What is

$$\begin{bmatrix} \frac{1}{2} & \frac{1}{3} & \frac{1}{4} \\ \frac{1}{5} & \frac{1}{6} & \frac{1}{7} \\ \frac{1}{8} & \frac{1}{9} & \frac{1}{10} \end{bmatrix} + \begin{bmatrix} 1 & 0 & 0 \\ 0 & 1 & 0 \\ 0 & 0 & 1 \end{bmatrix}?$$

3. What is

$$\begin{bmatrix} x & y & z \\ p & s & t \\ u & v & w \end{bmatrix} + \begin{bmatrix} 1-x & -y & -z \\ -p & 1-s & -t \\ -u & -v & 1-w \end{bmatrix}?$$

4. Does the sum

$$\begin{bmatrix} 3 & 2 & 1 \\ 1 & 3 & 2 \\ 3 & 1 & 2 \end{bmatrix} + 0_2$$

make sense? Does the sum

$$\begin{bmatrix} 3 & 2 & 1 \\ 1 & 3 & 2 \\ 3 & 1 & 2 \end{bmatrix} + 0_3$$

make sense? What is this latter sum?

1-5. Addition of Matrices Continued

Let A and B be two $n \times n$ matrices.

Then $A + B$ has been defined to be the matrix C whose entries are $[C]_{i,j} = [A]_{i,j} + [B]_{i,j}$. Similarly, $B + A$ is the matrix D whose entries are $[D]_{i,j} = [B]_{i,j} + [A]_{i,j}$. Now the entries of a matrix are just numbers. We know that the addition of numbers obeys the law $x + y = y + x$. Consequently,

$$[A]_{i,j} + [B]_{i,j} = [B]_{i,j} + [A]_{i,j}.$$

Thus, $[C]_{i,j} = [D]_{i,j}$. That is, every entry of the matrix C is the same as the corresponding entry of the matrix D. This means that the matrices C and D are equal. Since $C = A + B$, while $D = B + A$, we have proved the following theorem.

THEOREM

If A and B are both $n \times n$ matrices, then

$$A + B = B + A.$$

If we make more intelligent use of our notation for the entries of a matrix and make use of the short formula

$$[A + B]_{i,j} = [A]_{i,j} + [B]_{i,j}$$

defining the sum of two matrices, we can give the above proof in three lines, to wit

$$[A + B]_{i,j} = [A]_{i,j} + [B]_{i,j} = [B]_{i,j} + [A]_{i,j} = [B + A]_{i,j}. \quad (2)$$

Thus every entry of $A + B$ is equal to the corresponding entry in $B + A$, so that $A + B = B + A$. The student should be sure that he understands exactly why the equations (2) express exactly the same thing as the verbal proof which precedes it.

If we look back at the proof of this theorem, we see that we proved a theorem about the addition of matrices simply by using what we already knew about addition of numbers. Why was this possible? Because the addition of two matrices is defined simply by the addition of corresponding entries in the two matrices. If we keep this in mind, we will find that each of the basic laws for the addition of numbers leads to a corresponding law for the addition of matrices.

For instance, we have the following theorem.

THEOREM

If A, B, and C are all $n \times n$ matrices, then

$$(A + B) + C = A + (B + C).$$

PROOF

First we will give a proof using our notation for the entries in a matrix:

$$\begin{aligned} [(A + B) + C]_{i,j} = [A + B]_{i,j} + [C]_{i,j} &= \{[A]_{i,j} + [B]_{i,j}\} + [C]_{i,j} \\ &= [A]_{i,j} + \{[B]_{i,j} + [C]_{i,j}\} \\ &= [A]_{i,j} + [B + C]_{i,j} \\ &= [A + (B + C)]_{i,j}. \end{aligned}$$

Thus every entry of $(A + B) + C$ is equal to the corresponding entry of $A + (B + C)$, so that $(A + B) + C = A + (B + C)$.

Now let us express the very same proof verbally: The entry in the ith row and jth column of $A + B$ is, by definition, $[A]_{i,j} + [B]_{i,j}$. Thus, the entry in the ith row and jth column of $(A + B) + C$ is $\{[A]_{i,j} + [B]_{i,j}\} + [C]_{i,j}$. Similarly, the entry in the ith row and jth column of $B + C$ is $[B]_{i,j} + [C]_{i,j}$, so that the entry in the ith row and jth column of $A + (B + C)$ is $[A]_{i,j} + \{[B_{i,j}] + [C]_{i,j}\}$. Now, the entries of a matrix are merely numbers. For numbers, we know that

$$x + (y + z) = (x + y) + z.$$

Hence each entry of $A + (B + C)$ is the same as the corresponding entry of $(A + B) + C$. Consequently,

$$A + (B + C) = (A + B) + C. \qquad \text{Q.E.D.}$$

The theorem that has just been proved shows that the order in which matrices are added is immaterial. Consequently, we shall follow the ordinary practice of leaving out unnecessary parentheses and simply write

$$(A + B) + C = A + (B + C) = A + B + C.$$

Let A be an $n \times n$ matrix. Then to add the $n \times n$ zero matrix 0_n to A, we have simply to add the number zero to each of the entries of A. But addition of zero to an entry will not change the entry. Hence $A + 0_n$ has exactly the same entries as A. This proves the following theorem.

THEOREM

If A is an $n \times n$ matrix, then

$$A + 0_n = A.$$

Of course, since $A + 0_n = 0_n + A$, it is also true that

$$0_n + A = A.$$

Let us write the above theorem in a more abbreviated, more attractive way. If A is an $n \times n$ matrix, then the *only* zero matrix which can be added to A is the $n \times n$ zero matrix; we have simply not *defined* the addition of two matrices of different size. Hence, it is really not necessary to specify that the zero matrix in the last theorem is the $n \times n$ zero matrix—that is the only zero matrix it could possibly be. We can simply say the following.

THEOREM

If A is a matrix, then

$$A + 0 = 0 + A = 0.$$

Of course, in reading this equation, one should know that the matrix A will be of some particular size and that 0 means the zero matrix of the same size. But, knowing this, it is pointless to keep saying it over and over again. Whenever it is possible to drop the mention of a detail, it is better to do so than to let details multiply to the point where the main points get lost in a huge mass of detail.

Now we want to subtract matrices. This we do with the help of the following definitions.

DEFINITION

Let A be an $n \times n$ matrix. Then the negative matrix of A is the matrix each of whose entries is the negative of the corresponding entry of A.

The negative matrix of A will be denoted by the symbol $-A$. That is, the negative matrix $-A$ is defined by $[-A]_{i,j} = [A]_{i,j}$.

DEFINITION

If A and B are matrices of the same size, then the difference of A and B, denoted by $A - B$, is the sum of A and the negative of B. That is, $A - B$ is defined as $A + (-B)$.

We can easily prove the following theorems:

(*a*) $A + (-A) = 0$

(*b*) $-(-A) = A$

(*c*) $-0 = 0$

(*d*) $-(A + B) = (-A) + (-B).$

PROOF OF THE FIRST THEOREM

First we give a proof, using our notation for the entries in a matrix: By definition, $[-A]_{i,j} = -[A]_{i,j}$. Thus

$$[A + (-A)]_{i,j} = [A]_{i,j} + [-A]_{i,j} = [A]_{i,j} - [A]_{i,j} = 0,$$

so that $A + (-A) = 0$. Next let us express the same proof verbally: The entry in the ith row and jth column of $-A$ is, by definition, $-[A]_{i,j}$. Thus, the entry in the ith row and jth column of $A + (-A)$ is $[A]_{i,j} + (-[A]_{i,j})$. Hence, this entry is zero; that is, every entry of the matrix $A + (-A)$ is zero. That is, $A + (-A)$ is the zero matrix. Q.E.D.

The remaining theorems can be proved in the same way. The proofs will not be given, but will be left to the student as an exercise.

EXERCISES

1. Prove the second of the theorems stated immediately above.

2. Prove the third of the theorems stated immediately above.

3. Prove the fourth of the theorems stated immediately above.

4. Compute

$$\begin{bmatrix} 1 & 1 & 0 \\ 1 & 0 & 1 \\ \sqrt{2} & 1 & 0 \end{bmatrix} + \begin{bmatrix} 3 & 2 & 1 \\ 4 & 17 & 8 \\ 9 & 6 & 14 \end{bmatrix} - \begin{bmatrix} 0 & 1 & 3 \\ 14 & 8 & 6 \\ 1 - \sqrt{2} & 11 & 11 \end{bmatrix}.$$

5. Compute

$$\begin{bmatrix} \frac{1}{2} & \frac{1}{3} & \frac{1}{4} \\ \frac{1}{5} & \frac{1}{6} & \frac{1}{7} \\ \frac{1}{8} & \frac{1}{9} & \frac{1}{10} \end{bmatrix} - \begin{bmatrix} 0.3 & 0.4 & 0.5 \\ 0.6 & 0.7 & 0.8 \\ 0.9 & 1.0 & 1.1 \end{bmatrix}.$$

6. Compute

$$\begin{bmatrix} 1 & 2 & 3 \\ 4 & 5 & 6 \\ 7 & 8 & 9 \end{bmatrix} + \begin{bmatrix} 9 & 8 & 7 \\ 6 & 5 & 4 \\ 3 & 2 & 1 \end{bmatrix} - \begin{bmatrix} 0 & 0 & 0 \\ 0 & 0 & 0 \\ 0 & 0 & 0 \end{bmatrix} - \begin{bmatrix} 10 & 10 & 10 \\ 10 & 10 & 10 \\ 10 & 10 & 10 \end{bmatrix}.$$

(The teacher may supply other exercises of this same type.)

1-6. Addition of Matrices Concluded

The theorems given in Section 1-5 include exact analogs of all the basic laws of ordinary algebra, in so far as these laws refer to addition and subtraction. We know that all the more complicated algebraic laws about addition and subtraction are consequences of these basic laws. Since the basic laws for the addition and subtraction of matrices are the same as the basic laws for the addition and subtraction in ordinary algebra, all the other laws for the addition and subtraction of matrices must be the same as the corresponding laws for the addition and subtraction of numbers. That is:

In so far as only addition and subtraction are involved, the algebra of matrices is exactly like the ordinary algebra of numbers.

So you do not have to study the algebra of addition and subtraction of matrices—you already know it. But, the algebra which you already know is now filled with a new and much richer content. Formerly, it could be applied only to numbers. Now, it can be applied to matrices of arbitrary size. Thus, we obtained a very considerable result with a very small effort, simply by observing that our old algebraic laws of addition and subtraction apply not only to numbers, but also to quite different kinds of things, namely, matrices. This very powerful trick of pouring new wine into old bottles has been used many times, and often with great success, in the most modern mathematics.

A good example of the general principle emphasized above is provided by the following problem. Suppose that A and B are known matrices. How can we solve the equation

$$X + A = B$$

for the unknown matrix X? The answer is easy. We do exactly what we learned in ninth-year high school algebra. Add the matrix $-A$ to both sides. This gives

$$X + A + (-A) = B - A.$$

Since $A + (-A) = 0$, and $X + 0 = X$, we have

$$X = B - A.$$

This is our solution.

EXERCISES

1. Prove that, if A, B, and C are matrices all of the same size, $(A + C) - (A + B) = C - B$.

2. Solve the equation

$$X + \begin{bmatrix} 0 & 1 \\ 1 & 0 \end{bmatrix} = \begin{bmatrix} 1 & 0 \\ 0 & 1 \end{bmatrix}$$

for the matrix X.

3. Solve the equation

$$X + \begin{bmatrix} 0 & 0 & 1 \\ 0 & 1 & 0 \\ 1 & 0 & 0 \end{bmatrix} = \begin{bmatrix} 2 & 1 & 2 \\ 3 & 2 & 3 \\ 4 & 3 & 4 \end{bmatrix}$$

for the matrix X.

4. Solve the equation

$$X + \begin{bmatrix} 2 & 1 \\ 1 & -2 \end{bmatrix} = \begin{bmatrix} 3 & 0 \\ 1 & 3 \end{bmatrix}.$$

What is the solution of the equation

$$\begin{bmatrix} 2 & 1 \\ 1 & -2 \end{bmatrix} + X = \begin{bmatrix} 3 & 0 \\ 1 & 3 \end{bmatrix}?$$

5. Solve the equation

$$X - \begin{bmatrix} 1 & 1 \\ 1 & 9 \end{bmatrix} = \begin{bmatrix} 3 & 0 \\ 4 & 1 \end{bmatrix}$$

for the matrix X.

6. Solve the equation

$$X - \begin{bmatrix} 3 & 0 & 4 \\ 1 & 2 & 0 \\ 3 & 7 & 9 \end{bmatrix} = \begin{bmatrix} 1 & 2 & 3 \\ 7 & 1 & 9 \\ 8 & 1 & 2 \end{bmatrix} - \begin{bmatrix} 4 & 0 & 3 \\ 0 & 2 & 1 \\ 9 & 3 & 7 \end{bmatrix}$$

for the matrix X.

(The teacher may supply extra drill exercises of the same type.)

1-7. Numerical Multiples of Matrices

Once we know how to add numbers, it is customary to define $2x$ as the sum of $x + x$ or x with itself, $3x$ as the sum $2x$ and x, etc. Fractional parts of x are defined by requiring that $\frac{1}{2}x + \frac{1}{2}x = x$, $\frac{1}{3}x + \frac{1}{3}x + \frac{1}{3}x = x$, and so forth. All this can just as readily be done with matrices.

The sum $2A = A + A$ of two equal matrices is clearly the matrix each of whose entries is exactly twice the corresponding entry in A.

$$\begin{bmatrix} 2 & 3 \\ -1 & 0 \end{bmatrix} + \begin{bmatrix} 2 & 3 \\ -1 & 0 \end{bmatrix} = \begin{bmatrix} 2+2 & 3+3 \\ (-1)+(-1) & 0+0 \end{bmatrix} = \begin{bmatrix} 4 & 6 \\ -2 & 0 \end{bmatrix}.$$

The sum $3A = 2A + A = A + A + A$ is equally clearly the matrix each of whose entries is exactly three times the corresponding entry in A. For instance,

$$\begin{bmatrix} 2 & 3 \\ -1 & 0 \end{bmatrix} + \begin{bmatrix} 2 & 3 \\ -1 & 0 \end{bmatrix} + \begin{bmatrix} 2 & 3 \\ -1 & 0 \end{bmatrix} = \begin{bmatrix} 3 \times 2 & 3 \times 3 \\ 3 \times (-1) & 3 \times 0 \end{bmatrix}$$
$$= \begin{bmatrix} 6 & 9 \\ -3 & 0 \end{bmatrix}.$$

The equation $\frac{1}{2}A + \frac{1}{2}A = A$ defining the matrix $\frac{1}{2}A$ is clearly satisfied by the matrix each of whose entries is exactly one-half the corresponding entry of A; the equation $\frac{1}{3}A + \frac{1}{3}A + \frac{1}{3}A = A$ defining the matrix $\frac{1}{3}A$ is clearly satisfied by the matrix each of whose entries is exactly one-third the corresponding entry of A.

These considerations lead us to make the following general definition of the *numerical multiples of a given matrix A*.

DEFINITION

Let A be an $n \times n$ matrix and x any real number. Then the *product* of A *by the number* x is the matrix whose entry in the ith row and jth column is x times the corresponding entry $[A]_{i,j}$ in the matrix A. That is, the matrix xA is defined by the formula

$$[xA]_{i,j} = x[A]_{i,j}.$$

Note that we have defined numerical multiples of a matrix A, that is, the product of a matrix by a number, *not* the product of two matrices. It is possible to define the product of two matrices, but this is more complicated, and will be done later. We may now state and prove facts about the products of matrices by numbers.

THEOREM

If A and B are matrices of the same size, and x and y are numbers, then

(*a*) $x(yA) = (xy)A$
(*b*) $(x + y)A = (xA) + (yA)$
(*c*) $(-1)A = -A$
(*d*) $x(A + B) = xA + xB$
(*e*) $x \cdot 0 = 0$ (Any numerical multiple of the zero matrix is the zero matrix.)
(*f*) $0 \cdot A = 0$ (The product of any matrix by the number zero is the zero matrix.)

PROOF OF (*a*)

First let us give a proof using our notation for the entries in a matrix:

$$[x(yA)]_{i,j} = x[yA]_{i,j} = x\{y[A]_{i,j}\} = (xy)[A]_{i,j}$$
$$= [(xy)A]_{i,j}.$$

Now let us express this same proof verbally: The entry in the ith row and jth column of yA is, by definition, $y[A]_{i,j}$. The entry in the ith row and jth column of $x(yA)$ is consequently $x\{y[A]\}_{i,j}$. The entry in the ith row and jth column of $(xy)A$ is, by definition, $(xy)[A]_{i,j}$. Now, the entries in a matrix are numbers. For numbers, we know that $x(yz) = (xy)z$. Thus, each entry of the matrix $x(yA)$ is the same as the corresponding entry of the matrix $(xy)A$. Hence $x(yA) = (xy)A$. Q.E.D.

PROOF OF (*d*)

Using our notation for the entries in a matrix, we may simply write:

$$[x(A + B)]_{i,j} = x[(A + B)]_{i,j} = x\{[A]_{i,j} + [B]_{i,j}\}$$
$$= x[A]_{i,j} + x[B]_{i,j}$$
$$= [xA]_{i,j} + [xB]_{i,j}$$
$$= [(xA + xB)]_{i,j}.$$

This same proof may be expressed verbally, as follows. The entry in the ith row and jth column of the matrix $A + B$ is $[A]_{i,j} + [B]_{i,j}$. Thus, the entry in the ith row and jth column of the matrix $x(A + B)$ is $x\{[A]_{i,j} + [B]_{i,j}\}$. The entry in the ith row, jth column of the matrix xA is $x[A]_{i,j}$; that in the same row and column of the matrix xB is $x[B]_{i,j}$. Hence, the entry in the ith row, jth column of the matrix $xA + xB$ is $x[A]_{i,j} + x[B]_{i,j}$. Now, the entries in a matrix are numbers. For numbers, we know that $x(y + z) = xy + xz$. Thus each entry of the matrix $x(A + B)$ is the same as the corresponding entry of the matrix $xA + xB$. Hence $xA + xB = x(A + B)$. Q.E.D.

PROOF OF (f)

Using our notation for the entries in a matrix, we may simply write:

$$[0 \cdot A]_{i,j} = 0 \cdot [A]_{i,j} = 0.$$

Thus $0 \cdot A = 0$. The same proof may be expressed verbally, as follows. The entry in the ith row and jth column of the matrix $0 \cdot A$ is the product of the number zero and of the number $[A]_{i,j}$. But the product of zero by any number is zero. Hence, every entry in the matrix $0 \cdot A$ is zero. Thus $0 \cdot A$ is the zero matrix.

Parts (*b*), (*c*), and (*e*) of the above theorem may be proved in a similar way. They will not be proved, but will be left to the student to prove as an exercise.

When we studied the laws governing the addition and subtraction of matrices, we saw that they were the same as the laws governing addition and subtraction in ordinary algebra. Thus, without learning any new laws of algebra, we could simply apply our previous knowledge and apply the same old algebraic laws of addition and subtraction to matrices. The situation when we come to multiplication of matrices by numbers is *rather similar*, but *not exactly the same.* The various parts of the theorem that we have just proved resemble the basic algebraic laws for multiplication very closely. Thus, many of the more complicated algebraic laws and procedures governing ordinary multiplication still remain correct for expressions involving the multiplication of matrices by numbers. The difference is that the product of a number by a number is a number, but *the product of a matrix by a number is not a number but a matrix.*

Which laws and algebraic procedures remain valid for products of matrices by numbers, and which laws break down? We shall not try to answer this question by giving an exhaustive list. We shall simply say: with a little practice, it is not hard to guess. A guess can be checked by working out the algebra step by step, and then by trying to justify each step by using one of the parts (*a*) to (*e*) of the theorem proved above. Let us look at an example and see how this works.

Suppose we want to solve the matrix equation

$$-2\left(X + \begin{bmatrix} 1 & 2 & 3 \\ 0 & 1 & 2 \\ 0 & 0 & 1 \end{bmatrix}\right) = 3X + \begin{bmatrix} 1 & 0 & 0 \\ 0 & 0 & 0 \\ 0 & 0 & 1 \end{bmatrix}.$$

To solve equations of this general form in ordinary algebra, we simply multiply out, transpose, and divide. In matrix algebra, we do exactly the same; but if we want to be fussy and justify each step, we can, by doing one step at a time and by explicitly justifying each step by a reference to some one of our theorems on matrix algebra. In the case of the equation displayed above, our detailed steps (which really are just "multiply, transpose, divide" written out at great length) would be as follows: What we do first is multiply out [that is, use (*d*) of the above theorem] to get

$$-2X + \begin{bmatrix} -2 & -4 & -6 \\ 0 & -2 & -4 \\ 0 & 0 & -2 \end{bmatrix} = 3X + \begin{bmatrix} 1 & 0 & 0 \\ 0 & 0 & 0 \\ 0 & 0 & 1 \end{bmatrix}.$$

Then we add $+2X$ to both sides of the equation to get

$$\begin{bmatrix} -2 & -4 & -6 \\ 0 & -2 & -4 \\ 0 & 0 & -2 \end{bmatrix} = 3X + 2X + \begin{bmatrix} 1 & 0 & 0 \\ 0 & 0 & 0 \\ 0 & 0 & 1 \end{bmatrix}.$$

Then we use (*b*) of the theorem to find that $3X + 2X = 5X$, so that

$$\begin{bmatrix} -2 & -4 & -6 \\ 0 & -2 & -4 \\ 0 & 0 & -2 \end{bmatrix} = 5X + \begin{bmatrix} 1 & 0 & 0 \\ 0 & 0 & 0 \\ 0 & 0 & 1 \end{bmatrix}.$$

Adding

$$-\begin{bmatrix} 1 & 0 & 0 \\ 0 & 0 & 0 \\ 0 & 0 & 1 \end{bmatrix}$$

to both sides, we find that

$$\begin{bmatrix} -3 & -4 & -6 \\ 0 & -2 & -4 \\ 0 & 0 & -3 \end{bmatrix} = 5X.$$

Multiplying both sides of this last equation by $\frac{1}{5}$, we find [using (*a*) of the theorem to show that $\frac{1}{5}(5X) = X$]

$$X = \begin{bmatrix} -\frac{3}{5} & -\frac{4}{5} & -\frac{6}{5} \\ 0 & -\frac{2}{5} & -\frac{4}{5} \\ 0 & 0 & -\frac{3}{5} \end{bmatrix}.$$

This is our solution.

Of course, when one gets used to this, there is no reason to be so terribly fussy. One can simply say, as one does in ordinary algebra: multiply out, transpose, divide.

EXERCISES

1. Find values x, y, z, and a that satisfy the matrix relationship

$$\begin{bmatrix} x+3 & 2y-8 \\ z+1 & 4a-6 \end{bmatrix} = \begin{bmatrix} 0 & -6 \\ -3 & 2a \end{bmatrix}.$$

2. If

$$A = \begin{bmatrix} 3 & 2 & 1 \\ 4 & -5 & 6 \\ 0 & 8 & -3 \end{bmatrix}, \quad B = \begin{bmatrix} -3 & 4 & 8 \\ -2 & 6 & -1 \\ 0 & 2 & 3 \end{bmatrix}$$

determine the entry in the sum $A + B$ that is in

(*a*) The third row and second column
(*b*) The first row and third column
(*c*) The third row and first column.

3. Given

$$A = \begin{bmatrix} 1 & 0 & 2 \\ 3 & 1 & 4 \\ 5 & 0 & 6 \end{bmatrix}, \quad B = \begin{bmatrix} 2 & 1 & -1 \\ 3 & 0 & -2 \\ 0 & 1 & 1 \end{bmatrix}, \quad C = \begin{bmatrix} 4 & 0 & 2 \\ 1 & 0 & 0 \\ -2 & 4 & -4 \end{bmatrix}$$

compute the following:

(*a*) $A + B$ (*b*) $A + (B + C)$
(*c*) $A - B$ (*d*) $(A - B) + C$
(*e*) $(A + B) + C$ (*f*) $B - A$.

4. (*a*) In Exercise 3, consider the answers to parts (*b*) and (*e*). What law is illustrated?

(*b*) In Exercise 3, consider the answers to parts (*c*) and (*f*). What conclusions can be drawn?

5. If

$$\begin{bmatrix} 2 & -3 \\ 4 & 0 \end{bmatrix} - X = \begin{bmatrix} -3 & 4 \\ 5 & -1 \end{bmatrix}$$

determine the matrix X.

6. For

$$A = \begin{bmatrix} 2 & 2 & 2 \\ 2 & 1 & -3 \\ 1 & 0 & 4 \end{bmatrix}, \quad B = \begin{bmatrix} 3 & 3 & 3 \\ 3 & 0 & 5 \\ 6 & 9 & -1 \end{bmatrix},$$

$$C = \begin{bmatrix} 4 & 4 & 4 \\ 5 & -1 & 0 \\ 7 & 8 & -1 \end{bmatrix}$$

determine the result of the following operations:

(*a*) $2A - B + C$ (*b*) $3A - 4B - 2C$
(*c*) $7A - 2(B - C)$ (*d*) $3(A - 2B + 3C)$.

7. Assuming that the matrices are of the same size, prove that

$$D + (E + F) = (D + E) + F.$$

8. Prove part (*b*) of the theorem of Section 1-7.

9. Prove part (*c*) of the theorem of Section 1-7.

10. Prove part (*e*) of the theorem of Section 1-7.

11. Let A, B, and C be the matrices of Exercise 6. Solve the equation

$$\tfrac{1}{2}(X + A) = 3[X + (2X + B)] + C,$$

giving all the steps in detail and justifying each step.

12. Let A, B, and C be the matrices of Exercise 6. Solve the equations

(*a*) $2(X + B) = 3(X + (\frac{1}{2}X + A)) + C$
(*b*) $40(X + A) = 47(X + B) + 48(X + C)$
(*c*) $\sqrt{2}\,(X + C) = 31(X + \sqrt{2}\,(X + A - B))$

in as few steps as you dare.

(The teacher may invent and assign additional exercises like these.)

Chapter 2

MULTIPLICATION OF MATRICES

2-1. A Problem Arising in Business Management

By now, we have defined and studied

1. Addition of two matrices
2. Multiplication of a matrix by a number.

But we still have not defined the product of two matrices. To do this is our next aim. The formal definition is somewhat complicated and may at first seem odd. For this reason, it is well to show how a simple practical problem can lead us to combine two matrices in the way which we shall ultimately call multiplication:

Suppose that a manufacturer of hi-fi radio-phonograph sets produces four distinct models: *A*, *B*, *C*, *D*. These models contain the following number of subassemblies:

Model	*Amplifiers*	*Detectors*	*Speakers*	*Pickup arms*
A (the portable)...........	1	1	1	0
B (the table)..............	1	1	2	1
C (the console)...........	2	1	2	1
D (the de luxe)............	2	3	4	1

We can, if we like, arrange this table in the form of a matrix:

$$\text{Model}\begin{cases} A \\ B \\ C \\ D \end{cases} \overset{\overbrace{\text{Amplifiers} \quad \text{Detectors} \quad \text{Speakers} \quad \text{Pickup arms}}^{\text{Subassemblies}}}{\begin{bmatrix} 1 & 1 & 1 & 0 \\ 1 & 1 & 2 & 1 \\ 2 & 1 & 2 & 1 \\ 2 & 3 & 4 & 1 \end{bmatrix}}$$

Suppose now that each of the subassemblies contains the following number of basic parts:

	Tubes	*Condensers*	*Coils*	*Magnets*
Amplifier..........	8	3	4	0
Detector..........	4	6	3	0
Speaker...........	0	1	1	2
Pickup arm.......	0	1	1	1

Again, we can arrange our information in the form of a matrix:

$$\text{Subassembly}\begin{cases} \text{Amplifier} \\ \text{Detector} \\ \text{Speaker} \\ \text{Pickup arm} \end{cases} \overset{\overbrace{\text{Tubes} \quad \text{Condensers} \quad \text{Coils} \quad \text{Magnets}}^{\text{Parts}}}{\begin{bmatrix} 8 & 3 & 4 & 0 \\ 4 & 6 & 3 & 0 \\ 0 & 1 & 1 & 2 \\ 0 & 1 & 1 & 1 \end{bmatrix}}$$

The first of the above matrices might be called the *models-subassemblies-requirement matrix;* the second might be called the *subassemblies–parts-requirement matrix.*

Now, it is perfectly clear that, by combining these two sets of information, we can easily compute the number of parts of each kind required for any model. That is, by correctly combining the entries in the *models–subassemblies-requirement matrix* with the entries in the *subassemblies–parts-requirement matrix*, we can figure out a *models–parts-requirement matrix*. How is this to be done? The answer is obvious. For instance, to compute the number of coils required for model C, we figure as follows:

We need	*Number of coils*
2 amplifiers with 4 coils each.............	8
1 detector with 3 coils....................	3
2 speakers with 1 coil each...............	2
1 pickup arm with 1 coil.................	1
Total..............................	14

Thus, each model C set produced requires 14 coils. How, in terms of our requirements matrices, did we get this answer for the *number of coils* in *model C?* We multiplied each entry in the *model C row* of the first matrix (reading from left to right) by the corresponding element in the *coils column* of the second matrix (reading from top to bottom) and then added all the resulting products.

It is clear that this procedure is perfectly general. Thus, to find the number of condensers required for model D, we multiply each element in the *model D row* of the first matrix by the corresponding element in the *condensers column* of the second matrix and then add, getting a total of

$$(2 \times 3) + (3 \times 6) + (4 \times 1) + (1 \times 1) = 29 \text{ condensers.}$$

In general, to get the element in the x row and y column of the models–parts-requirements matrix, we multiply each of the four elements in the x row of the first matrix by the element in the y column of the second matrix and add the four resulting products.

It is clear that this rule is a purely mathematical one, so that it can be adapted to give a rule for combining any two 4×4 matrices A and B: the "combination matrix" will be the matrix whose entry in the ith row, jth column is determined by multiplying each of the four elements in the ith row of the first matrix A by the corresponding element in the jth column of the second matrix B and then adding all four products. Thus, the entry in the ith row and jth column of the combination matrix will be

$$[A]_{i,1}[B]_{1,j} + [A]_{i,2}[B]_{2,j} + [A]_{i,3}[B]_{3,j} + [A]_{i,4}[B]_{4,j}.$$

There is, moreover, no reason why this rule should be restricted to 4×4 matrices. For instance, if A and B are 3×3 matrices, we can take their combination matrix to be the 3×3 matrix whose entry in the ith row, jth column, is determined by multiplying each of the three elements in the ith row of the first matrix A by the corresponding element in the jth column of the second matrix B and then adding all three products. Thus, the entry in the ith row and jth column of the 3×3 "combination matrix" will be

$$[A]_{i,1}[B]_{1,j} + [A]_{i,2}[B]_{2,j} + [A]_{i,3}[B]_{3,j}.$$

The mathematical rule for combining matrices to which we have been led is commonly called *multiplication of matrices.* But now it is time to give a formal, general definition.

EXERCISES

Combine the following matrices according to the rule developed in the preceding section:

1. $$\begin{bmatrix} 1 & 2 \\ 3 & 4 \end{bmatrix} \quad \text{and} \quad \begin{bmatrix} 5 & 6 \\ 7 & 8 \end{bmatrix}$$

2. $$\begin{bmatrix} 1 & 2 & 3 \\ 4 & 5 & 6 \\ 7 & 8 & 9 \end{bmatrix} \quad \text{and} \quad \begin{bmatrix} 1 & 0 & 1 \\ 1 & 1 & -1 \\ -1 & 0 & 1 \end{bmatrix}$$

3. $$\begin{bmatrix} 1 & 2 & -1 & -2 \\ 3 & -4 & 4 & -3 \\ 5 & -6 & 7 & -8 \\ 8 & -1 & 0 & 0 \end{bmatrix} \quad \text{and} \quad \begin{bmatrix} \frac{1}{2} & \frac{1}{3} & \frac{1}{4} & \frac{1}{5} \\ \frac{1}{2} & -\frac{1}{3} & -\frac{1}{4} & -\frac{1}{5} \\ 0 & 0 & 0 & 0 \\ \frac{1}{7} & \frac{1}{8} & \frac{1}{9} & \frac{1}{10} \end{bmatrix}.$$

(The teacher may invent and assign additional exercises of this type.)

2-2. Formal Definition of Multiplication

Now we shall give a formal definition of the product of two matrices and thereby bring our matrix algebra to a much higher stage of usefulness and perfection.

DEFINITION

Let A and B be the two $n \times n$ matrices. Then the product AB is the matrix whose entry in the ith row and jth column is

$$[A]_{i,1}[B]_{1,j} + [A]_{i,2}[B]_{2,j} + [A]_{i,3}[B]_{3,j} + \cdots + [A]_{i,n}[B]_{n,j}.$$

Thus, using our notation for the entries in a matrix, we may write the definition of the product matrix AB of two $n \times n$ matrices as

$$[AB]_{i,j} = [A]_{i,1}[B]_{1,j} + [A]_{i,2}[B]_{2,j} + \cdots + [A]_{i,n}[B]_{n,j}.$$

This definition is very useful, but unfortunately seems rather complicated at first. Let us take some time to become familiar with it.

The definition tells us: to find the entry in the ith row and jth column of the product AB,

Multiply the element in the ith row, *first* column of A by the element in the *first* row, jth column of B

Multiply the element in the ith row, *second* column of A by the element in the *second* row, jth column of B

Multiply the element in the ith row, *third* column of A by the element in the *third* row, jth column of B

. . . .

Multiply the element in the ith row, nth column of A by the element in the nth row, jth column of B

and then add all the terms obtained.

Suppose, for instance, that A and B are 2×2 matrices:

$$A = \begin{bmatrix} 1 & 2 \\ 3 & 4 \end{bmatrix} \qquad B = \begin{bmatrix} 5 & 6 \\ 7 & 8 \end{bmatrix}.$$

Then the definition tells us how to determine each of the entries of AB, as follows:

$$\begin{aligned}
[AB]_{1,1} &= [A]_{1,1}[B]_{1,1} + [A]_{1,2}[B]_{2,1} = 1 \times 5 + 2 \times 7 \\
&= 5 + 14 = 19 \\
[AB]_{1,2} &= [A]_{1,1}[B]_{1,2} + [A]_{1,2}[B]_{2,2} = 1 \times 6 + 2 \times 8 \\
&= 6 + 16 = 22 \\
[AB]_{2,1} &= [A]_{2,1}[B]_{1,1} + [A]_{2,2}[B]_{2,1} = 3 \times 5 + 4 \times 7 \\
&= 15 + 28 = 43 \\
[AB]_{2,2} &= [A]_{2,1}[B]_{1,2} + [A]_{2,2}[B]_{2,2} = 3 \times 6 + 4 \times 8 \\
&= 18 + 32 = 50.
\end{aligned}$$

Thus

$$AB = \begin{bmatrix} 19 & 22 \\ 43 & 50 \end{bmatrix}.$$

The rule for multiplication is *multiply row by column and then add;* that is, to find the element in the ith row and jth column of the product matrix AB, *multiply each element in the ith row of A by the corresponding element in the jth column of B and add all the resulting terms.*

The simplest way to carry this out in practice is by the following geometric scheme. Given two matrices A and B to be multiplied, *first* write A and then write B immediately above and to the right of A, in the following position:

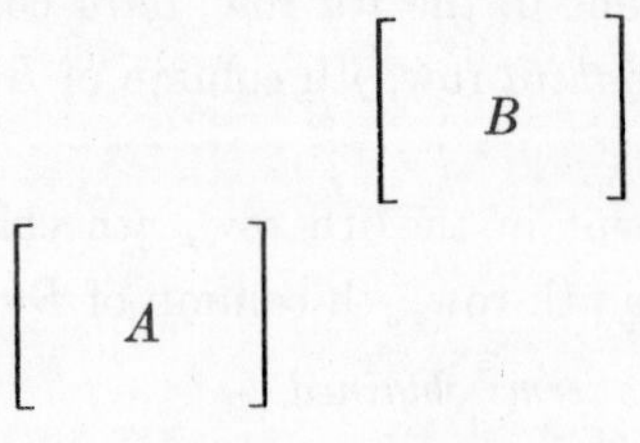

Then, to the right of A and below B draw a pair of brackets inside which the "answer," that is, the entries of AB, is to be written:

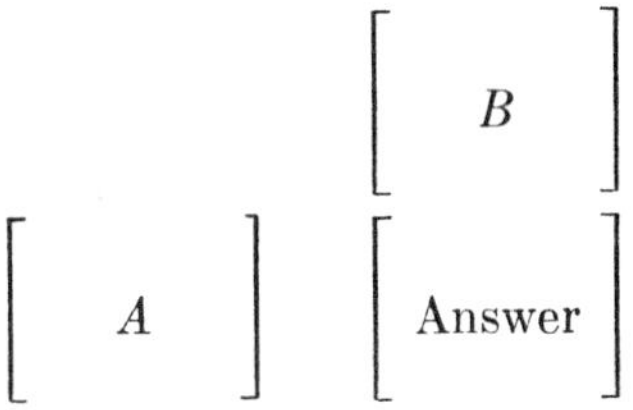

The ith row of A and the jth column of B will, together, point at some definite spot within the answer bracket; this is where the entry $[AB]_{i,j}$ should be written.

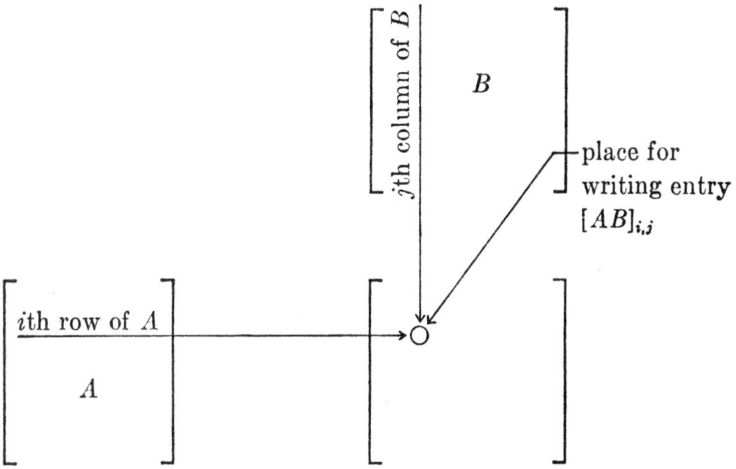

To determine the entry $[AB]_{i,j}$, we multiply each element of the row that points at it (starting from the left) by the corresponding element of the column that points at it (starting from the top) and add all the terms we get. Thus, to multiply

$$A = \begin{bmatrix} 1 & 2 & 3 \\ 4 & 5 & 6 \\ 7 & 8 & 9 \end{bmatrix} \quad \text{by} \quad B = \begin{bmatrix} 1 & 0 & 1 \\ 2 & 1 & 2 \\ 4 & 1 & 3 \end{bmatrix}$$

we first write

$$\begin{matrix} & \begin{bmatrix} 1 & 0 & 1 \\ 2 & 1 & 2 \\ 4 & 1 & 3 \end{bmatrix} \\ \begin{bmatrix} 1 & 2 & 3 \\ 4 & 5 & 6 \\ 7 & 8 & 9 \end{bmatrix} & \begin{bmatrix} & & \\ & & \\ & & \end{bmatrix} \end{matrix}$$

To determine the entry in the first row and first column of the product A, B, we would compute as follows:

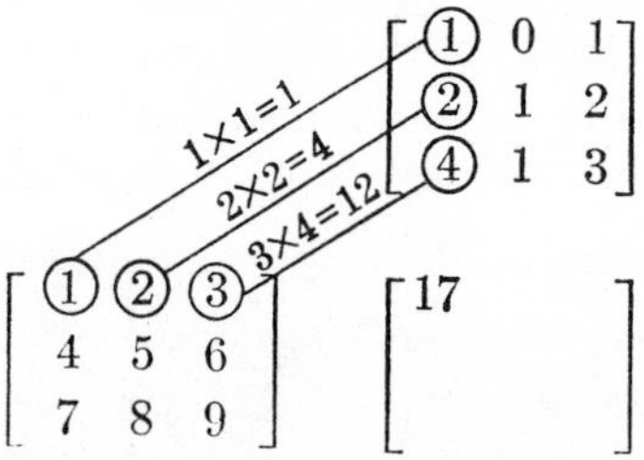

$$1 + 4 + 12 = 17$$

With a little practice, it is not hard to do all this mentally. Determining one entry of the product after another in this way, we finally would obtain the complete answer for the product AB:

$$\begin{matrix} & \begin{bmatrix} 1 & 0 & 1 \\ 2 & 1 & 2 \\ 4 & 1 & 3 \end{bmatrix} \\ \begin{bmatrix} 1 & 2 & 3 \\ 4 & 5 & 6 \\ 7 & 8 & 9 \end{bmatrix} & \begin{bmatrix} 17 & 5 & 14 \\ 38 & 11 & 32 \\ 59 & 17 & 50 \end{bmatrix} \end{matrix}.$$

(Check each of the entries of the answer yourself.) That is,

$$AB = \begin{bmatrix} 17 & 5 & 14 \\ 38 & 11 & 32 \\ 59 & 17 & 50 \end{bmatrix}.$$

(To get the answer, 27 multiplications and 9 additions were necessary.)

The definition of the product of two $n \times n$ matrices can be expressed in terms of the Σ notation for sums. Let us recall that in the Σ notation we write the sum

$$s = x_1 + \cdots + x_n$$

of n numbers as

$$s = \sum_{k=1}^{n} x_k.$$

In this notation the sum

$$[A]_{i,1}[B]_{1,j} + [A]_{i,2}[B]_{2,j} + \cdots + [A]_{i,n}[B]_{n,j}$$

would consequently be written as

$$\sum_{k=1}^{n} [A]_{i,k}[B]_{k,j}$$

so that we could say:

The product AB of two $n \times n$ matrices is the matrix whose entry in the ith row and jth column is

$$\sum_{k=1}^{n} [A]_{i,k}[B]_{k,j}.$$

If we make somewhat heavier use of our notation for the entries in a matrix, we may write even more compactly

$$[AB]_{i,j} = \sum_{k=1}^{n} [A]_{i,k}[B]_{k,j}.$$

Note that we have defined the product of two matrices *only if they are of the same size.* Two matrices will never be multiplied (or added) if they are of different sizes. We will not even define a rule by which matrices of different sizes could be multiplied.

EXERCISES

1. Let

$$A = \begin{bmatrix} 1 & 2 & 3 \\ 4 & 5 & 6 \\ 7 & 8 & 9 \end{bmatrix}, \quad B = \begin{bmatrix} 1 & 0 & -1 \\ -1 & 0 & 1 \\ -2 & 0 & -1 \end{bmatrix}, \quad C = \begin{bmatrix} \frac{1}{2} & \frac{1}{3} & \frac{1}{4} \\ \frac{1}{5} & \frac{1}{6} & \frac{1}{7} \\ \frac{1}{8} & \frac{1}{9} & \frac{1}{10} \end{bmatrix}.$$

Compute

(*a*) AB (*b*) AC
(*c*) BC (*d*) $(AB)C$
(*e*) $A(BC)$ (*f*) $A(B + C)$
(*g*) $A(B - C)$ (*h*) $AB - AC$
(*i*) $A \cdot A - B \cdot B + C \cdot C$ (*j*) $(A \cdot A) \cdot A$
(*k*) $((A \cdot A) \cdot A)A$ (*l*) $(((A \cdot A) \cdot A) \cdot A) \cdot A$.

2. Perform the following matrix multiplications:

(*a*)
$$\begin{bmatrix} 1 & 0 \\ 0 & 1 \end{bmatrix} \begin{bmatrix} 6 & 4 \\ 2 & 0 \end{bmatrix}$$

(*b*)
$$\begin{bmatrix} 1 & 0 & 0 \\ 0 & 1 & 0 \\ 0 & 0 & 1 \end{bmatrix} \begin{bmatrix} x_1 & x_2 & x_3 \\ y_1 & y_2 & y_3 \\ z_1 & z_2 & z_3 \end{bmatrix}$$

(*c*)
$$\begin{bmatrix} r_1 & r_2 & r_3 \\ s_1 & s_2 & s_3 \\ t_1 & t_2 & t_3 \end{bmatrix} \begin{bmatrix} 2 & 0 & 0 \\ 0 & 2 & 0 \\ 0 & 0 & 2 \end{bmatrix}$$

(*d*)
$$\begin{bmatrix} \alpha_1 & 0 & 0 \\ 0 & \alpha_2 & 0 \\ 0 & 0 & \alpha_3 \end{bmatrix} \begin{bmatrix} a_1 & a_2 & a_3 \\ b_1 & b_2 & b_3 \\ c_1 & c_2 & c_3 \end{bmatrix}$$

(*e*)
$$\begin{bmatrix} 0 & 0 & 0 \\ 0 & b_1 & 0 \\ 0 & 0 & 0 \end{bmatrix} \begin{bmatrix} x_1 & x_2 & x_3 \\ y_1 & y_2 & y_3 \\ z_1 & z_2 & z_3 \end{bmatrix}.$$

3. If

$$A = \begin{bmatrix} 1 & -1 \\ 0 & 2 \end{bmatrix}, \quad B = \begin{bmatrix} 1 & -1 \\ 4 & 0 \end{bmatrix}, \quad \text{and } C = \begin{bmatrix} 1 & 2 \\ 3 & 4 \end{bmatrix}$$

test the rule that $(AB)C = A(BC)$.

4. Let I denote the matrix

$$\begin{bmatrix} 1 & 0 & 0 \\ 0 & 1 & 0 \\ 0 & 0 & 1 \end{bmatrix}.$$

Let A, B, C be as in Exercise 1. Compute

(a) AI (b) BI
(c) CI (d) $(AI)I$
(e) $((AI)I)I$.

2-3. A Surprising Property of Matrix Multiplication

Let

$$A = \begin{bmatrix} 0 & 0 \\ 1 & 0 \end{bmatrix} \quad \text{and} \quad B = \begin{bmatrix} 0 & 1 \\ 0 & 0 \end{bmatrix}.$$

If we compute AB, we find

$$\begin{array}{cc} & \begin{bmatrix} 0 & 1 \\ 0 & 0 \end{bmatrix} \\ \begin{bmatrix} 0 & 0 \\ 1 & 0 \end{bmatrix} & \begin{bmatrix} 0 & 0 \\ 0 & 1 \end{bmatrix}. \end{array}$$

Thus

$$AB = \begin{bmatrix} 0 & 0 \\ 0 & 1 \end{bmatrix}.$$

If we compute BA, we find

$$\begin{array}{cc} & \begin{bmatrix} 0 & 0 \\ 1 & 0 \end{bmatrix} \\ \begin{bmatrix} 0 & 1 \\ 0 & 0 \end{bmatrix} & \begin{bmatrix} 1 & 0 \\ 0 & 0 \end{bmatrix}. \end{array}$$

Thus

$$BA = \begin{bmatrix} 1 & 0 \\ 0 & 0 \end{bmatrix}.$$

Thus AB and BA are different matrices. That is:

The product AB of two matrices can be entirely different from the product BA of the same two matrices taken in the reverse order.

Thus, *in writing products of matrices, we must pay careful attention to the order of the factors.* We must, consequently, distinguish between the result of multiplying B on the right by A to get BA

and the result of multiplying B on the left by A to get AB. In the algebra of numbers, these two operations of "right multiplication" and "left multiplication" are the same; in matrix algebra, they are entirely different.

This is a first difference between matrix algebra and ordinary algebra—and a very significant difference it is indeed. When we multiply numbers, we can use the law $xy = yz$ to rearrange factors in any way we like. When we multiply matrices, we have no such law and must, consequently, be careful to take the factors in the order given and not rearrange them carelessly.

Another way of putting this is: *multiplication of two matrices is not commutative.*

EXERCISES

1. Let

$$A = \begin{bmatrix} 1 & 2 \\ 3 & 4 \end{bmatrix} \qquad B = \begin{bmatrix} 1 & 0 \\ -1 & 1 \end{bmatrix}.$$

Calculate

(*a*) AB (*b*) BA
(*c*) $(AB)A$ (*d*) $(BA)A$
(*e*) $(BA)B$ (*f*) $B(BA)$
(*g*) $A(AB)$ (*h*) $((BA)A)B$
(*i*) $((AB)A)B$.

2. Let

$$A = \begin{bmatrix} 1 & 2 & 3 \\ 4 & 5 & 6 \\ 7 & 8 & 9 \end{bmatrix} \quad \text{and} \quad B = \begin{bmatrix} 1 & 0 & -1 \\ -1 & 0 & 1 \\ 1 & 1 & -1 \end{bmatrix}.$$

Calculate

(*a*) AB (*b*) BA
(*c*) $(AB)A$ (*d*) $(BA)A$
(*e*) $(BA)B$ (*f*) $B(BA)$
(*g*) $A(AB)$ (*h*) $((BA)A)B$
(*i*) $((AB)A)B$.

3. Let A and B be as in Exercise 2, and let

$$I = \begin{bmatrix} 1 & 0 & 0 \\ 0 & 1 & 0 \\ 0 & 0 & 1 \end{bmatrix}.$$

Calculate

(a) AI (b) IA
(c) BI (d) IB
(e) $(AB)I$ (f) $(AI)B$
(g) $I(AB)$.

2-4. Matrix Multiplication Continued

Let

$$A = \begin{bmatrix} 0 & 0 \\ 1 & 0 \end{bmatrix} \qquad B = \begin{bmatrix} 0 & 0 \\ 2 & 0 \end{bmatrix}.$$

Then $A \neq 0$ and $B \neq 0$. But if we compute AB, we find

$$\begin{array}{cc} & \begin{bmatrix} 0 & 0 \\ 2 & 0 \end{bmatrix} \\ \begin{bmatrix} 0 & 0 \\ 1 & 0 \end{bmatrix} & \begin{bmatrix} 0 & 0 \\ 0 & 0 \end{bmatrix} \end{array}$$

so that $AB = 0$. Thus:

The product of two matrices can be zero without either of the two matrices being zero.

This is a second major difference between ordinary algebra and matrix algebra.

The breakdown of the law that $xy = yx$ and of the law that $xy = 0$ only if either x or y is zero for matrix algebra causes many other things to go wrong. For instance: we know that a real number a can have at most two square roots, that is, that there are at most two roots of the equation $x \cdot x = a$. What is the proof? Simply this: suppose that $y \cdot y = a$ also. Then

Step 1: $x \cdot x = y \cdot y$
Step 2: $x \cdot x - y \cdot y = 0$
Step 3: $(x - y)(x + y) = x \cdot x + (-y \cdot x + x \cdot y) - y \cdot y$
Step 4: $y \cdot x = x \cdot y$
Step 5: From steps 3 and 4, $(x - y)(x + y) = x \cdot x - y \cdot y$
Step 6: From steps 5 and 2, $(x - y)(x + y) = 0$
Step 7: Therefore either $x - y = 0$ or $x + y = 0$
Step 8: Therefore either $x = y$ or $x = -y$.

For matrices, step 4 fails. But even if step 4 did not fail, step 7 would fail. So the proof is completely wrong if we try to apply it to matrices. In fact, it is false that a matrix can have at most two square roots. We have

$$\begin{bmatrix} 1 & 0 \\ 0 & 1 \end{bmatrix} \begin{bmatrix} 1 & 0 \\ 0 & 1 \end{bmatrix} = \begin{bmatrix} 1 & 0 \\ 0 & 1 \end{bmatrix}$$

$$\begin{bmatrix} 1 & 0 \\ 0 & -1 \end{bmatrix} \begin{bmatrix} 1 & 0 \\ 0 & -1 \end{bmatrix} = \begin{bmatrix} 1 & 0 \\ 0 & 1 \end{bmatrix}$$

$$\begin{bmatrix} -1 & 0 \\ 0 & 1 \end{bmatrix} \begin{bmatrix} -1 & 0 \\ 0 & 1 \end{bmatrix} = \begin{bmatrix} 1 & 0 \\ 0 & 1 \end{bmatrix}$$

$$\begin{bmatrix} -1 & 0 \\ 0 & -1 \end{bmatrix} \begin{bmatrix} -1 & 0 \\ 0 & -1 \end{bmatrix} = \begin{bmatrix} 1 & 0 \\ 0 & 1 \end{bmatrix}.$$

(Check all the multiplications yourself.)
Thus the matrix

$$I = \begin{bmatrix} 1 & 0 \\ 0 & 1 \end{bmatrix}$$

has the *four* different square roots

$$I = \begin{bmatrix} 1 & 0 \\ 0 & 1 \end{bmatrix}, \quad J = \begin{bmatrix} 1 & 0 \\ 0 & -1 \end{bmatrix}, \quad K = \begin{bmatrix} -1 & 0 \\ 0 & 1 \end{bmatrix},$$

$$L = \begin{bmatrix} -1 & 0 \\ 0 & -1 \end{bmatrix}.$$

Nor is this all. Given any number x, we have

$$\begin{bmatrix} 0 & x \\ \frac{1}{x} & 0 \end{bmatrix}\begin{bmatrix} 0 & x \\ \frac{1}{x} & 0 \end{bmatrix} = \begin{bmatrix} 1 & 0 \\ 0 & 1 \end{bmatrix}.$$

(Check the multiplication.) By giving x any one of an infinite number of different real values, we obtain an infinite number of different square roots of the matrix I:

$$\begin{bmatrix} 0 & 2 \\ \frac{1}{2} & 0 \end{bmatrix}, \quad \begin{bmatrix} 0 & 3 \\ \frac{1}{3} & 0 \end{bmatrix}, \quad \begin{bmatrix} 0 & 4 \\ \frac{1}{4} & 0 \end{bmatrix}, \quad \text{etc.}$$

Thus the very simple 2×2 matrix I has infinitely many distinct square roots.

This should make it plain that the fact that a *number* has at most two square roots is by no means trivial.

EXERCISES

1. Let

$$A = \begin{bmatrix} 1 & -1 \\ 0 & 2 \end{bmatrix} \qquad B = \begin{bmatrix} 1 & 0 \\ 1 & 2 \end{bmatrix}.$$

Show by computation that

(*a*) $(A + B)(A + B) \neq A^2 + 2AB + B^2$

(*b*) $(A + B)(A - B) \neq A^2 - B^2$.

2. Find at least eight square roots of the matrix

$$\begin{bmatrix} 1 & 0 & 0 \\ 0 & 1 & 0 \\ 0 & 0 & 1 \end{bmatrix}.$$

How many more square roots can you find?

3. Show that the matrix

$$A = \begin{bmatrix} 0 & 0 \\ 1 & 0 \end{bmatrix}$$

satisfies $A^2 = 0$. How many 2×2 matrices satisfying this equation can you find?

4. Show that the matrix

$$A = \begin{bmatrix} 0 & 0 & 0 \\ 1 & 0 & 0 \\ 0 & 1 & 0 \end{bmatrix}$$

satisfies $A(A \cdot A) = 0$. What is $A \cdot A$?

5. Let A, B, and I be the matrices

$$A = \begin{bmatrix} 1 & 2 & 3 \\ 4 & 5 & 6 \\ 7 & 8 & 9 \end{bmatrix}$$

$$B = \begin{bmatrix} 0 & \frac{1}{2} & \frac{1}{3} \\ 1 & 0 & 1 \\ 2 & 3 & 2 \end{bmatrix}$$

$$I = \begin{bmatrix} 1 & 0 & 0 \\ 0 & 1 & 0 \\ 0 & 0 & 1 \end{bmatrix}.$$

Compute

(*a*) AI (*b*) $(AI)I$
(*c*) $A(I \cdot I)$ (*d*) $((AI)I)I$
(*e*) $(IA)I$ (*f*) $((IA)I)I$
(*g*) $((I(IA))I)I$ (*h*) (BI)
(*i*) $(BI)I$ (*j*) $B(II)$
(*k*) $((BI)I)I$ (*l*) $(IB)I$
(*m*) $((IB)I)I$ (*n*) $((I(IB))I)I$.

What does this make you suspect about the effect of multiplying by I? Can you prove it?

2-5. The Unit Matrix

The matrix

$$I_2 = \begin{bmatrix} 1 & 0 \\ 0 & 1 \end{bmatrix}$$

is called the unit matrix of size 2. The matrix

$$I_3 = \begin{bmatrix} 1 & 0 & 0 \\ 0 & 1 & 0 \\ 0 & 0 & 1 \end{bmatrix}$$

is called the unit matrix of size 3. The matrix

$$I_4 = \begin{bmatrix} 1 & 0 & 0 & 0 \\ 0 & 1 & 0 & 0 \\ 0 & 0 & 1 & 0 \\ 0 & 0 & 0 & 1 \end{bmatrix}$$

is called the unit matrix of size 4, and so forth. In general, the unit matrix of size n, which we may denote by the symbol I_n, is the matrix whose entries are

$$[I_n]_{i,j} = 0 \qquad \text{if } i \neq j.$$
$$[I_n]_{i,j} = 1 \qquad \text{if } i = j, \text{ that is, } [I]_{j,j} = 1 \text{ for every } j.$$

Since only matrices of the same size will ever be added or multiplied, it will often be unnecessary for us to specify the size of the unit matrix we are talking about. If this is the case, we may leave off the subscript of the unit matrix and write it simply as I. This will be done only when the reader knows what size matrix is meant, or when it does not matter what size matrix is meant. Hence, it should never lead to confusion.

The reason for the special importance of the unit matrix is stated in the following theorem:

THEOREM

Let A be an $n \times n$ matrix. Then

$$AI = A \qquad \text{and} \qquad IA = A.$$

Thus, the unit matrix plays the same role in matrix algebra that the number 1 does in ordinary algebra: multiplication of any matrix A by the unit matrix (either on the left or on the right) just gives A back again.

PROOF

By definition of multiplication, the entry in the ith row and jth column of the product AI is

$$[A]_{i,1}[I]_{1,j} + [A]_{i,2}[I]_{2,j} + [A]_{i,3}[I]_{3,j} + \cdots + [A]_{i,n}[I]_{n,j}.$$

Since $[I]_{k,j} = 0$ whenever k is different from j, every term but one in this last expression is equal to zero and drops out, and we are left with just one term: $[A]_{i,j}[I]_{j,j}$. Since $[I]_{j,j} = 1$, $[A]_{i,j}[I]_{j,j} = [A]_{i,j}$. Thus, the entry in the ith row and jth column of the product AI is simply $A_{i,j}$. Thus AI and A have exactly the same entries, and therefore $AI = A$.

The fact that $IA = A$ may be proved in exactly the same way. The proof will not be given, but is left to the student as an exercise. Q.E.D.

EXERCISE

Prove that the equation $IA = A$ holds for every matrix A.

2-6. The Laws of Matrix Multiplication

We saw a little while ago that two basic laws which govern multiplication in the algebra of ordinary numbers break down when it comes to matrices. That is, we are faced with the

Breakdown of the commutative law: The product AB of two matrices may be entirely different from the product BA of the same two matrices.

Breakdown of the law of cancellation: The product AB of two matrices may be zero even if both factors A and B are other than zero.

At this point one may fear a total collapse of all the other familiar laws of algebra and a boundless algebraic chaos ensuing. But such is not the case. Aside from the commutative and cancellation laws, most of the other basic laws of ordinary algebra remain valid for matrices. These laws are stated in the various parts of the following theorem.

THEOREM

Let A, B, and C be $n \times n$ matrices. Then

(*a*) $(AB)C = A(BC)$ (associative law)
(*b*) $A(B + C) = AB + AC$ (distributive law)
(*c*) $(B + C)A = BA + CA$ (distributive law)

PROOF OF PART (*b*)

Using our notation for the entries of a matrix, we may give the proof rather simply, as follows:

$$\begin{aligned}[A(B + C)]_{i,j} &= [A]_{i,1}[B + C]_{1,j} + [A]_{i,2}[B + C]_{2,j} + \cdots \\ &\qquad + [A]_{i,n}[B + C]_{n,j} \\ &= [A]_{i,1}\{[B]_{1,j} + [C]_{1,j}\} + [A]_{i,2}\{[B]_{2,j} + [C]_{2,j}\} \\ &\qquad + \cdots + [A]_{i,n}\{[B]_{n,j} + [C]_{n,j}\}. \\ [A(B + C)]_{i,j} &= \{[A]_{i,1}[B]_{1,j} + [A]_{i,2}[B]_{2,j} + \cdots + [A]_{i,n}[B]_{n,j}\} \\ &\quad + \{[A]_{i,1}[C]_{1,j} + [A]_{i,2}[C]_{2,j} + \cdots + [A]_{i,n}[B]_{n,j}\} \\ &= [AB]_{i,j} + [AC]_{i,j} \\ &= [AB + AC]_{i,j}.\end{aligned}$$

The same proof may be expressed verbally, as follows: Let $D = A(B + C)$. The entry in the kth row and jth column of the sum matrix $B + C$ is $[B]_{k,j} + [C]_{k,j}$. The entry in the ith row and jth column of $A(B + C)$ is consequently

$$\begin{aligned}[D]_{i,j} = [A]_{i,1}\{[B]_{1,j} + [C]_{1,j}\} \\ + [A]_{i,2}\{[B]_{2,j} + [C]_{2,j}\} \\ + [A]_{i,3}\{[B]_{3,j} + [C]_{3,j}\} + \cdots \\ + [A]_{i,n}\{[B]_{n,j} + [C]_{n,j}\}.\end{aligned}$$

Since the entries in a matrix are ordinary numbers, we can use ordinary algebra to multiply out all the terms of this expression and rearrange, getting

$$\begin{aligned}[D]_{i,j} = {} & \{[A]_{i,1}[B]_{1,2} + [A]_{i,2}[B]_{2,j} + \cdots + [A]_{i,n}[B]_{n,j}\} \\ & + \{[A]_{i,1}[C]_{1,j} + [A]_{i,2}[C]_{2,j} + \cdots + [A]_{i,n}[C]_{n,j}\}.\end{aligned}$$

The sum in the first set of braces may now be recognized as the sum defining the entry $[AB]_{i,j}$ of the product matrix AB, and the sum in the second set of braces may be recognized as the sum defining the entry $[AC]_{i,j}$ of the product matrix AC. Thus

$$[D]_{i,j} = [AB]_{i,j} + [AC]_{i,j}.$$

Hence each entry of D is identical with the corresponding entry of the sum $AB + AC$. Therefore, $D = AB + AC$. Since D was defined as $D = A(B + C)$, it follows that

$$A(B + C) = AB + AC. \qquad \text{Q.E.D.}$$

Part (*c*) of the above theorem may be proved in the same way by writing down the entries of $(B + C)A$ and using ordinary algebra to rearrange terms. The proof will not be given, but will be left as an exercise.

It should be noted that, if the commutative law $AB = BA$ held for matrices, it would be entirely unnecessary to prove parts (*b*) and (*c*) of the above theorem separately, since the two statements

$$A(B + C) = AB + AC$$

and

$$(B + C)A = BA + CA$$

would say exactly the same thing. But since for matrices "multiplication on-the left" and "multiplication on the right" are quite distinct, the two formulae above say different things, even though both are true.

Part (*a*) of the above theorem may be proved in a similar way by "expanding and rearranging." But since the details are somewhat complicated, the proof will be omitted.

Part (*a*) of our theorem permits us to omit parentheses in

expressions such as $(AB)C$ and to follow the ordinary algebraic convention of writing

$$(AB)C = A(BC) = ABC.$$

Parts (*b*) and (*c*) of our theorem justify "multiplying out," *but only if all the factors of a product are kept in correct order.* The formulae

$$A(B + C) = AB + AC$$
$$(B + C)A = BA + BC$$

are always correct for matrices; but *formulae like*

$$A(B + C) = AB + CA$$
$$A(B + C) = BA + CA,$$

etc., *can be entirely false.*

Matrix algebra and ordinary numerical algebra may be compared as follows: The two are very much the same in so far as only addition and subtraction are involved. The ordinary laws for multiplying out remain valid for matrices, but only if the factors of a product are kept in the correct order. Since the commutative law fails for matrices, right-hand multiplication and left-hand multiplication must be distinguished; products cannot be rearranged in different order, and careful attention must be paid to the order of factors. The product of two matrices can be zero even if neither factor is zero.

EXERCISES

1. Prove that if A, B, and C are $n \times n$ matrices, then

$$(B + C)A = BA + CA.$$

2. Let

$$A = \begin{bmatrix} 0 & 1 \\ 1 & 0 \end{bmatrix} \quad B = \begin{bmatrix} 0 & 1 \\ -1 & 0 \end{bmatrix} \quad C = \begin{bmatrix} 1 & 0 \\ 0 & 0 \end{bmatrix}.$$

Test the formulae

$$\begin{aligned} A(B + C) &= AB + AC \\ (B + C)A &= BA + CA \\ A(B + C) &= AB + CA \\ A(B + C) &= BA + CA; \end{aligned}$$

which are correct, and which are false?

3. Let

$$A = \begin{bmatrix} 0 & 0 \\ 1 & 0 \end{bmatrix} \qquad B = \begin{bmatrix} 1 & 0 \\ 0 & 0 \end{bmatrix}.$$

Show that $AB \neq 0$, but $BA = 0$.

4. Show that for all

$$A = \begin{bmatrix} a & b \\ -b & a \end{bmatrix} \qquad B = \begin{bmatrix} c & d \\ -d & c \end{bmatrix}$$

$$AB = BA.$$

Illustrate by assigning appropriate values to a, b, c, and d. (a,b,c,d integers.)

5. Show that for all

$$A = \begin{bmatrix} ab & b^2 \\ -a^2 & -ab \end{bmatrix}$$

$$A \cdot A = 0.$$

Illustrate by assigning appropriate values to a and b. (a, b integers.)

6. Let

$$D = \begin{bmatrix} 0 & 1 \\ 1 & 0 \end{bmatrix} \qquad E = \begin{bmatrix} 0 & -1 \\ 1 & 0 \end{bmatrix} \qquad F = \begin{bmatrix} 1 & 0 \\ 0 & -1 \end{bmatrix}.$$

Compute the following:

(*a*) DE (*b*) DF
(*c*) EF (*d*) ED
(*e*) FD (*f*) FE.

If $AB = -BA$, A and B are said to be anticommutative. What conclusions can be drawn concerning D, E, F?

7. Show that the matrix

$$A = \begin{bmatrix} 3 & 1 \\ -1 & 2 \end{bmatrix}$$

is a solution of the equation $A \cdot A - 5A + 7I = 0$.

8. Find the value of x which makes the following product equal I:

$$\begin{bmatrix} 2 & 0 & 7 \\ 0 & 1 & 0 \\ 1 & 2 & 1 \end{bmatrix} \begin{bmatrix} -x & -14x & 7x \\ 0 & 1 & 0 \\ x & 4x & -2x \end{bmatrix}.$$

9. Let

$$A = \begin{bmatrix} 0 & 0 & 0 \\ 1 & 0 & 0 \\ 0 & 1 & 0 \end{bmatrix} \quad B = \begin{bmatrix} 0 & 0 & 0 \\ 0 & 0 & 0 \\ 1 & 0 & 0 \end{bmatrix} \quad C = \begin{bmatrix} 0 & 0 & 0 \\ 2 & 0 & 0 \\ 1 & 2 & 0 \end{bmatrix}.$$

Show that $AB = BA$, $AC = CA$, $BC = CB$.

Let

$$D = \begin{bmatrix} 0 & 1 & 0 \\ 0 & 0 & 1 \\ 0 & 0 & 0 \end{bmatrix}.$$

Show that $AD \neq DA$, $BD \neq DB$, $CD \neq DC$.

2-7. Powers of Matrices. Laws of Exponents

There is sufficient similarity between the laws of matrix multiplication and the ordinary laws of numerical multiplication for us to define the *powers of a matrix* in the ordinary way and find the ordinary law of exponents.

We define $A^2 = A \cdot A$, $A^3 = A \cdot A^2$, $A^4 = A \cdot A^3$, etc. That is, we make the *inductive definition:* $A^2 = A \cdot A$, $A^{n+1} = A \cdot A^n$. It is also convenient to take $A^1 = A$ and $A^0 = I$.

The following theorem gives the law of exponents.

THEOREM

Let A be a matrix. Then

$$A^n \cdot A^m = A^{n+m}.$$

PROOF

We will give a proof by mathematical induction. If $n = 1$, then

$$A^1 \cdot A^m = A^{1+m}$$

says exactly the same thing as

$$A \cdot A^m = A^{m+1}$$

and is true by definition.

Next, suppose that we know our theorem to be true for a given value k of n, i.e., suppose that

$$A^k \cdot A^m = A^{k+m}. \tag{1}$$

We will show that the theorem is true for the next value $k + 1$ of n. This we do as follows: since $A^{k+1} = A \cdot A^k$ by definition, we have

$$A^{k+1} \cdot A^m = (A \cdot A^k) \cdot A^m.$$

Then, using the associative law,

$$A^{k+1}A^m = A(A^k \cdot A^m).$$

It follows by our inductive assumption (1) that

$$A^{k+1} \cdot A^m = A \cdot A^{k+m}.$$

Thus, by definition,

$$A^{k+1}A^m = A^{k+m+1}.$$

Since $k + m + 1 = (k + 1) + m$.

$$A^{k+1}A^m = A^{k+1}A^m.$$

Consequently, our theorem follows by the principle of mathematical induction. Q.E.D.

It should be observed that the law of exponents applies only to powers of a single matrix A. *We cannot conclude that*

$$ABAB^2 = A^2B^3,$$

since this would involve rearrangement of factors. *Such a formula can very well be false.* Care must be taken to keep the factors in a product in correct order.

On the other hand, a formula like

$$ABA^2AA^4B^2A = ABA^7B^2A,$$

which involves only the law of exponents and not the rearrangement of factors, is necessarily true.

EXERCISES

1. Explain why in matrix algebra $(A + B)(A - B) \neq A^2 - B^2$ except in special cases. Can you devise two matrices A and B which will illustrate the inequality? Can you devise two matrices A and B which will illustrate the special case?

(*Hint:* Use square matrices, size 2.)

2. Let

$$A = \begin{bmatrix} 1 & 2 \\ 3 & 4 \end{bmatrix} \qquad B = \begin{bmatrix} 1 & -1 \\ 0 & 2 \end{bmatrix}.$$

Compute

(*a*) AB (*b*) BA
(*c*) ABA (*d*) ABB
(*e*) BAB (*f*) $ABBA$
(*g*) $BABA$ (*h*) BAA.

Which of these matrices are the same? If A and B were not matrices but numbers, which expressions would be the same?

3. Let

$$A = \begin{bmatrix} 1 & 2 & 3 \\ -1 & 1 & 1 \\ 4 & 5 & 0 \end{bmatrix}.$$

Compute A^2, A^3, A^4, and A^5.

(The teacher may invent and assign additional exercises of this kind.)

2-8. Multiplication of Matrices by Matrices and Multiplication of Matrices by Numbers

We have introduced two quite different notions of multiplication in connection with matrices. One was the notion of multiplying two matrices A and B of the same size to get a product AB. The other was the notion of multiplying a matrix A by a number x to produce a numerical multiple xA of the matrix A.

(Let us recall that the numerical multiple xA was defined, in Chapter 1, as the matrix each of whose entries was exactly x times the corresponding entry in the matrix A.)

There are a number of important relations between these two types of multiplication. These are stated in the following theorem.

THEOREM

Let A and B be matrices of the same size, and let x be a real number. Then

$$x(AB) = (xA)B = A(xB).$$

Note that this theorem permits free rearrangement of the *numerical* factors in a product. For instance,

$$(2A)(3A) = 2((3A)A) = 2(3(A \cdot A)) = 6A^2;$$
$$(2A)(3B) = 2((3A) \cdot B) = 2(3(AB)) = 6AB.$$

Of course, $6AB$ must be distinguished from $6BA$; only the *numerical* factors, and not the *matrix* factors, of a product may be rearranged.

The above theorem, the law of exponents in Section 2-7, the laws of multiplication in Section 2-6, the properties of the unit matrix given in Section 2-5, and the laws of addition and subtraction and of multiplication by numerical factors given in Chapter 1, when used in combination, permit us to carry over many of the familiar manipulations of elementary algebra to the algebra of matrices.

Let us take a simple example: "multiply out" the expression $(A + 2B)^3$. Now

$$\begin{aligned}(A + 2B)^3 &= (A + 2B)((A + 2B)(A + 2B)) \\ &= (A + 2B)(A(A + 2B) + 2B(A + 2B)) \\ &= (A + 2B)((A^2 + 2AB) + (2BA + 4B^2)) \\ &= A(A^2 + 2AB + 2BA + 4B^2) \\ &\qquad + 2B(A^2 + 2AB + 2BA + 4B^2) \\ &= A^3 + 2A^2B + 2ABA + 4AB^2 \\ &\qquad + 2BA^2 + 4BAB + 4B^2A + 8B^3.\end{aligned}$$

And there we must stop. If A and B were numbers, we could add $2A^2B + 2ABA + 2BA^2$ to get $6A^2B$, and $4AB^2 + 4BAB + 4B^2A$ to get $12AB^2$. But this involves a rearrangement of factors which is permissible for numbers, impermissible for matrices.

PROOF OF THEOREM (OPTIONAL)

Using our notation for the entries in a matrix, we may write the proof of the statement $(xA)B = x(AB)$ as follows:

$$\begin{aligned}[(xA)B]_{i,j} &= [xA]_{i,1}[B]_{1,j} + [xA]_{i,2}[B]_{2,j} + \cdots + [xA]_{i,n}[B]_{n,j} \\ &= \{x[A]_{i,1}\}[B]_{1,j} + \{x[A]_{i,2}\}[B]_{2,j} + \cdots \\ &\qquad + \{x[A]_{i,n}\}[B]_{n,j} \\ &= x\{[A]_{i,1}[B]_{1,j} + [A]_{i,2}[B]_{2,j} + \cdots + [A]_{i,n}[B]_{n,j}\} \\ &= x[AB]_{i,j} = [x(AB)]_{i,j}.\end{aligned}$$

The same proof may be expressed verbally, as follows: let $D = (xA)B$. The entry in the ith row and kth column of xA is $x[A]_{i,k}$. Thus, the entry in the ith row and jth column of the product matrix $D = (xA)B$ is

$$\begin{aligned}[D]_{i,j} = \{x[A]_{i,1}\}[B]_{1,j} + \{x[A]_{i,2}\}[B]_{2,j} + \{x[A]_{i,3}\}[B]_{3,j} + \cdots \\ + \{x[A]_{i,n}\}[B]_{n,j}.\end{aligned}$$

Since the entries in a matrix are numbers, we can use ordinary algebra to rewrite the expression on the right-hand side, obtaining

$$[D]_{i,j} = x\{[A]_{i,1}[B]_{1,j} + [A]_{i,2}[B]_{2,j} + \cdots + [A]_{i,n}[B]_{n,j}\}.$$

The expression inside the braces may now be recognized as the entry $[AB]_{i,j}$ in the ith row and jth column of the product matrix AB. Thus

$$[D]_{i,j} = x[AB]_{i,j}.$$

Thus D and $x(AB)$ have the same entries, so that $D = x(AB)$. Since we defined D by $D = (xA)B$, it follows that

$$(xA)B = x(AB).$$

The fact that $A(xB) = x(AB)$ may be proved in a similar way. The proof will not be given, but will be left to the student as an exercise. Q.E.D.

The theorem proved in this section permits us to omit parentheses in expressions like $A(xB)$ and $(xA)B$, and write in the ordinary algebraic fashion simply AxB and xAB, etc.

EXERCISES

1. Prove that the binomial theorem can be used to expand the expression $(A + bI)^n$, A being a matrix and b a number. Prove that this theorem cannot be used to expand the expression $(A + B)^n$, A and B being two matrices of the same size.

What do you get on "expanding" $(A + B)^2$? On expanding $(A + B)^3$?

2. Multiply out:

(*a*) $(A - 2B)^2(A + C)$

(*b*) $(A - B + C)^3(A - C)$

(*c*) $(A - \frac{1}{2}B)^4$.

3. Show that the matrix

$$A = \begin{bmatrix} 1 & -1 \\ -1 & 1 \end{bmatrix}$$

satisfies the equation $A^2 - 2A = 0$.

4. Show that the matrix

$$A = \begin{bmatrix} 2 & 3 \\ 3 & 2 \end{bmatrix}$$

satisfies the equation $A^2 - 4A - 5I = 0$.

5. Show that the matrix

$$A = \begin{bmatrix} 1 & 0 & 0 \\ 2 & 1 & 0 \\ 3 & 2 & 1 \end{bmatrix}$$

satisfies the equation $A^3 - 3A^2 + 3A + I = 0$.

6. Show that the matrix

$$A = \begin{bmatrix} 2 & 3y \\ \frac{3}{y} & 2 \end{bmatrix}$$

satisfies the equation $A^2 - 4A - 5I = 0$ no matter what the value of y and that this quadratic equation consequently has an infinite number of distinct 2×2 matrices as roots.

7. Prove the parts of the theorem of the preceding section which are not proved in the text.

2-9. Polynomials in a Matrix

We have seen that the law of exponents is the same for powers of a matrix as for powers of a number and that *numerical factors* in a product may be rearranged at will. Moreover, we know that the laws governing addition and subtraction of matrices are just like the laws governing addition and subtraction of numbers. It follows from all this that the algebraic rules for manipulating expressions which are made up out of sums of powers of a *single matrix* A, multiplied by arbitrary *numerical coefficients,* are just like the algebraic rules for manipulating ordinary polynomials. Thus, for instance,

$$\begin{aligned}(2A + I)^2 &= 4A \cdot 4A + 2A \cdot I + I \cdot 2A + I \cdot I \\ &= 16A^2 + 4A + I,\end{aligned}$$

and

$$\begin{aligned}(A - I)(2A + I)^2 &= (A - I)(16A^2 + 4A + I)\\ &= 16A^3 + 4A^2 + A - 16A^2 - 4A - I\\ &= 16A^3 - 12A^2 - 3A - I,\end{aligned}$$

etc. The most general expression of this form can be written as

$$c_mA^m + c_{m-1}A^{m-1} + \cdots + c_0I, \tag{2}$$

that is, as the sum of powers of the single matrix A multiplied by the numerical coefficients $c_m, c_{m-1}, \ldots, c_0$. Such an expression (2) will be called a polynomial in the matrix A.

It should be noted that we reserve this term for the expression (2) in which the coefficients $c_m, \ldots, c_0$ are *numbers*. If we tried to allow the coefficients $c_m, \ldots, c_0$ to be matrices, the fact that matrix multiplication does not obey the commutative law would introduce many complications. By restricting our attention to the case of numerical coefficients, we are able to avoid at least some of those complications.

An equation of the form

$$c_mA^m + c_{m-1}A^{m-1} + \cdots + c_0I = 0$$

with numerical coefficients $c_m, c_{m-1}, \ldots, c_0$ will be called a *polynomial equation satisfied by the matrix* A. We shall see in the next chapter that many important properties of a matrix can be discovered by studying the polynomial equations which it satisfies.

EXERCISES

1. Show that

$$A^n - I = (A - I)(A^{n-1} + A^{n-2} + \cdots + I).$$

2. Factor the following polynomials in A into a product of first-degree polynomials in A:

(*a*) $A^2 - 5A + 6I$

(*b*) $A^3 - 6A^2 + 12A - 8I$

(*c*) $A^3 + 2A^2 - A + 2I$.

3. Find a matrix satisfying the polynomial equations

(*a*) $X^2 - 5X + 6I = 0$

(*b*) $X^3 - 6X^2 + 12X - 8I = 0$

(*c*) $X^3 + 2X^2 + X + 2I = 0.$

How many matrices satisfying each of these polynomial equations can you find?

Chapter 3

DIVISION OF MATRICES

3-1. Introduction

In the first two chapters of this book, we have studied addition, subtraction, and multiplication of matrices. It should not be surprising that our next aim is to study the division of matrices.

Division, in algebra, is the problem of solving the equation

$$AX = B$$

for the unknown (matrix) X. Of course, since matrix multiplication is not commutative, we will have to expect that the solution X of this equation can be different from the solution Y of the equation

$$YA = B.$$

That is, since right-hand and left-hand multiplication are different, we must expect right-hand and left-hand division to be different also.

Thus, for instance, the equation

$$\begin{bmatrix} 1 & 0 \\ 2 & 1 \end{bmatrix} X = \begin{bmatrix} 1 & 2 \\ 1 & 2 \end{bmatrix}$$

has the solution

$$X = \begin{bmatrix} 1 & 2 \\ -1 & -2 \end{bmatrix},$$

while the equation

$$X\begin{bmatrix}1 & 0\\ 2 & 1\end{bmatrix} = \begin{bmatrix}1 & 2\\ 1 & 2\end{bmatrix}$$

has the solution

$$Y = \begin{bmatrix}-3 & 2\\ -3 & 2\end{bmatrix}.$$

(Check these solutions by multiplying out.) This shows vividly that right- and left-hand division must be distinguished.

But actually we have an even more fundamental difficulty than this.

Suppose, for instance, that we try to solve the equation

$$\begin{bmatrix}0 & 0\\ 1 & 0\end{bmatrix} X = \begin{bmatrix}1 & 2\\ 3 & 4\end{bmatrix}. \tag{1}$$

Suppose that X is the matrix

$$X = \begin{bmatrix}a & b\\ c & d\end{bmatrix}.$$

Then

$$\begin{bmatrix}0 & 0\\ 1 & 0\end{bmatrix} X = \begin{bmatrix}0 & 0\\ a & b\end{bmatrix}.$$

Thus, no matter what entries we try to take for the matrix X, the matrix

$$\begin{bmatrix}0 & 0\\ 1 & 0\end{bmatrix} X$$

will *always* have zero everywhere in its first row. The equation consequently has no solution. Thus, division by the matrix

$$A = \begin{bmatrix}0 & 0\\ 1 & 0\end{bmatrix}$$

is just as impossible as "division by zero" in ordinary algebra.

How, then, can we hope to solve the equation

$$AX = B?$$

Ordinary algebra suggests the following line of reasoning: If the matrix A had a *reciprocal matrix* A^{-1} *satisfying* $A^{-1}A = I$, then we could multiply both sides of the equation $AX = B$ on the left by the matrix A^{-1} and would find

$$A^{-1}B = A^{-1}AX = (A^{-1}A)X = IX = X,$$

since $A^{-1}A = I$. Thus the equation $AX = B$ would have the solution $X = A^{-1}B$, and the situation would be just as it is in ordinary algebra.

Now, *some* matrices A do have reciprocal matrices A^{-1} satisfying $A^{-1}A = I$. For instance, it is easy to check that

$$\begin{bmatrix} 0 & \frac{1}{2} \\ \frac{1}{2} & 0 \end{bmatrix} \cdot \begin{bmatrix} 0 & 2 \\ 2 & 0 \end{bmatrix} = \begin{bmatrix} 1 & 0 \\ 0 & 1 \end{bmatrix}.$$

Thus the matrix

$$A = \begin{bmatrix} 0 & 2 \\ 2 & 0 \end{bmatrix}$$

has the reciprocal matrix

$$\begin{bmatrix} 0 & \frac{1}{2} \\ \frac{1}{2} & 0 \end{bmatrix}.$$

The equation

$$\begin{bmatrix} 0 & 2 \\ 2 & 0 \end{bmatrix} X = \begin{bmatrix} 1 & 2 \\ 3 & 4 \end{bmatrix}$$

may consequently be solved by multiplying both sides by the reciprocal matrix to find

$$X = \begin{bmatrix} 0 & \frac{1}{2} \\ \frac{1}{2} & 0 \end{bmatrix} \begin{bmatrix} 1 & 2 \\ 3 & 4 \end{bmatrix} = \begin{bmatrix} \frac{3}{2} & 2 \\ \frac{1}{2} & 1 \end{bmatrix}.$$

The matrix

$$\begin{bmatrix} 0 & 0 \\ 1 & 0 \end{bmatrix}$$

on the other hand, can have no reciprocal matrix; for if it did, we could solve the equation

$$\begin{bmatrix} 0 & 0 \\ 1 & 0 \end{bmatrix} X = \begin{bmatrix} 1 & 2 \\ 3 & 4 \end{bmatrix}$$

by multiplying both sides by the reciprocal matrix, and we have seen above that this equation has no solution.

Summarizing: The problem of division of matrices may be reduced to that of finding the reciprocal matrix of a given matrix. Some matrices have reciprocal matrices, some do not.

We consequently wish to answer the following questions:

(*a*) Which matrices have reciprocal matrices, and which do not?

(*b*) If a matrix A has a reciprocal matrix, how can we calculate this reciprocal matrix?

This chapter is devoted to studying these two questions. To answer them is not easy; before answering them, we shall have to study a number of other basic properties of matrices.

EXERCISES

1. Show by multiplying out that if $ad - bc \neq 0$, the 2×2 matrix

$$A = \begin{bmatrix} a & b \\ c & d \end{bmatrix}$$

has the reciprocal

$$A^{-1} = \begin{bmatrix} \dfrac{d}{ad - bc} & \dfrac{-b}{ad - bc} \\ \dfrac{-c}{ad - bc} & \dfrac{a}{ad - bc} \end{bmatrix}.$$

2. Using Exercise 7, find reciprocals of the following matrices:

(*a*) $A_1 = \begin{bmatrix} 2 & 1 \\ 4 & 3 \end{bmatrix}$ (*b*) $A_2 = \begin{bmatrix} 5 & 3 \\ -1 & 4 \end{bmatrix}$

(*c*) $A_3 = \begin{bmatrix} -3 & 5 \\ -2 & 7 \end{bmatrix}$ (*d*) $A_4 = \begin{bmatrix} -1 & 9 \\ -2 & 8 \end{bmatrix}$

(*e*) $A_5 = \begin{bmatrix} 3 & -1 \\ 5 & -2 \end{bmatrix}$ (*f*) $A_6 = \begin{bmatrix} -2 & 0 \\ 3 & 4 \end{bmatrix}$

(*g*) $A_7 = \begin{bmatrix} 3 & 1 \\ -2 & 1 \end{bmatrix}$ (*h*) $A_8 = \begin{bmatrix} 5 & 2 \\ -1 & 6 \end{bmatrix}.$

3. Using Exercise 2, solve the following equations for the matrix X:

(*a*) $\begin{bmatrix} 2 & 1 \\ 4 & 3 \end{bmatrix} X = \begin{bmatrix} 18 & -2 \\ -12 & 6 \end{bmatrix}$ (*b*) $\begin{bmatrix} 5 & 3 \\ -1 & 4 \end{bmatrix} X = \begin{bmatrix} 0 & 23 \\ 0 & 0 \end{bmatrix}$

(*c*) $\begin{bmatrix} -3 & 5 \\ -2 & 7 \end{bmatrix} X = \begin{bmatrix} -11 & 0 \\ 0 & 11 \end{bmatrix}$ (*d*) $\begin{bmatrix} -1 & 9 \\ -2 & 8 \end{bmatrix} X = \begin{bmatrix} 3 & 1 \\ 0 & 1 \end{bmatrix}$

(*e*) $\begin{bmatrix} 3 & -1 \\ 5 & -2 \end{bmatrix} X = \begin{bmatrix} 1 & 2 \\ 3 & 4 \end{bmatrix}$ (*f*) $\begin{bmatrix} -2 & 0 \\ 3 & 4 \end{bmatrix} X = \begin{bmatrix} 2 & 4 \\ 6 & 8 \end{bmatrix}$

(*g*) $\begin{bmatrix} 3 & 1 \\ -2 & 1 \end{bmatrix} X = \begin{bmatrix} 105 & 15 \\ 0 & 0 \end{bmatrix}$ (*h*) $\begin{bmatrix} 5 & 2 \\ -1 & 6 \end{bmatrix} X = \begin{bmatrix} 0 & 1 \\ 0 & 1 \end{bmatrix}.$

4. Show that the matrix

$$A = \begin{bmatrix} 0 & 1 & 0 \\ 0 & 0 & 1 \\ 1 & 0 & 0 \end{bmatrix}$$

has the reciprocal matrix

$$A^{-1} = \begin{bmatrix} 0 & 0 & 1 \\ 1 & 0 & 0 \\ 0 & 1 & 0 \end{bmatrix}.$$

Show that $A^{-1}A = I$ and that $AA^{-1} = I$ also. Solve the equations

$$AX = \begin{bmatrix} 1 & 2 & 3 \\ 4 & 5 & 6 \\ 7 & 8 & 9 \end{bmatrix}$$

$$YA = \begin{bmatrix} 1 & 2 & 3 \\ 4 & 5 & 6 \\ 7 & 8 & 9 \end{bmatrix}.$$

Do left-hand division and right-hand division really have to be distinguished?

5. Show that if a matrix A satisfies any one of the equations

$$A^2 = 0, \qquad A^3 = 0, \qquad A^4 = 0, \qquad \text{etc.,}$$

it can have no reciprocal matrix. Use this fact to show that the matrix

$$\begin{bmatrix} 0 & 0 & 0 & 0 \\ 1 & 0 & 0 & 0 \\ 2 & 3 & 0 & 0 \\ 4 & 5 & 6 & 0 \end{bmatrix}$$

has no reciprocal matrix.

3-2. Using an Equation Satisfied by a Matrix

Let us take a 2×2 matrix like

$$A = \begin{bmatrix} 1 & 2 \\ 3 & 4 \end{bmatrix}.$$

If we calculate A^2, we find that

$$A^2 = \begin{bmatrix} 7 & 10 \\ 15 & 22 \end{bmatrix}.$$

If we subtract I from A and $7I$ from A^2, we get a pair of matrices each with the entry zero in the upper left-hand corner:

$$A - I = \begin{bmatrix} 0 & 2 \\ 3 & 3 \end{bmatrix}$$

$$A^2 - 7I = \begin{bmatrix} 0 & 10 \\ 15 & 15 \end{bmatrix}.$$

If we subtract five times the first of these two matrices from the second of these two matrices, we are bound to get a matrix with zero everywhere in the first row; actually making the subtraction, we get zero for every entry of the difference:

$$(A^2 - 7I) - 5(A - I) = \begin{bmatrix} 0 & 0 \\ 0 & 0 \end{bmatrix}.$$

Thus the matrix A satisfies the equation

$$A^2 - 5A - 2I = 0.$$

Once we have discovered this equation, we can use it to find a reciprocal for the matrix A. We have only to write the equation as

$$(A - 5I)A = 2I$$

or

$$(\tfrac{1}{2}(A - 5I))A = I.$$

This makes it plain that the matrix $\frac{1}{2}(A - 5I)$ is a reciprocal of A. Thus A has the reciprocal

$$A^{-1} = \tfrac{1}{2}\begin{bmatrix} -4 & 2 \\ 3 & -1 \end{bmatrix} = \begin{bmatrix} -2 & 1 \\ \frac{3}{2} & -\frac{1}{2} \end{bmatrix}.$$

This reasoning shows that to find a reciprocal matrix of A is easy once we have found a polynomial equation satisfied by A.

The polynomial equation satisfied by A is useful in doing other calculations involving A also. Suppose, for instance, that we want to calculate A raised to a high power, say, to the eighth power. If we did this in the most obvious way, multiplying A by itself over and over again eight times, we would have a lot of arithmetic to do. Each time we multiply two 2×2 matrices, we have eight numerical multiplications and four numerical additions to perform. Thus, to raise A to the eighth power by repeated multiplication requires $8 \times 8 = 64$ multiplications and $4 \times 8 = 32$ additions.

Using the equation satisfied by the matrix A, we can cut down the amount of arithmetic required drastically. We can write our equation as

$$A^2 = 5A + 2I.$$

Squaring both sides, we find

$$\begin{aligned} A^4 = (5A + 2I)^2 &= 25A^2 + 20A + 4I \\ &= 25(5A + 2I) + 20A + 4I \\ &= 145A + 54I. \end{aligned}$$

Squaring both sides again, we find

$$\begin{aligned} A^8 = (145A + 54I)^2 &= 21{,}025A^2 + 15{,}660A + 2{,}916I \\ &= 21{,}025(5A + 2I) + 15{,}660A + 2{,}916I \\ &= 120{,}785A + 44{,}966I. \end{aligned}$$

Thus

$$\begin{aligned} A^8 &= 120{,}785 \begin{bmatrix} 1 & 2 \\ 3 & 4 \end{bmatrix} + 44{,}966 \begin{bmatrix} 1 & 0 \\ 0 & 1 \end{bmatrix} \\ &= \begin{bmatrix} 165{,}751 & 241{,}570 \\ 362{,}355 & 528{,}106 \end{bmatrix}. \end{aligned}$$

Try working this out by repeated multiplication, and see if you can carry it through to the end.

The above considerations show us that once we are able to find a polynomial equation satisfied by a matrix A, it is much easier to study all the other properties of the matrix.

For this reason, the following theorem is very important.

MAIN THEOREM

Let A be an $n \times n$ matrix. Then A satisfies a polynomial equation of the form

$$A^k + c_{k-1}A^{k-1} + \cdots + c_0 I = 0$$

of a degree k which is not more than n and with numerical coefficients $c_{k-1}, \ldots, c_0$.

Thus, a 2×2 matrix always satisfies either a linear or a quadratic polynomial equation. A 3×3 matrix always satisfies a cubic, quadratic, or linear equation. A 4×4 matrix always satisfies a polynomial equation of degree 4 at most, etc.

This theorem will provide us with the key to the problem of reciprocal matrices.

Unfortunately, the proof of the above theorem is too complicated to be given in this book.

EXERCISES

1. Find a quadratic equation satisfied by the matrix

$$A = \begin{bmatrix} 1 & 2 \\ -2 & -1 \end{bmatrix}.$$

Use the equation to find the reciprocal matrix of A.

2. Find the seventh power of the matrix A of Exercise 1.

3. Find a quadratic equation satisfied by the matrix

$$A = \begin{bmatrix} 1 & -2 \\ 7 & -14 \end{bmatrix}.$$

Does this matrix have a reciprocal matrix?

4. Find a quadratic equation satisfied by the matrix

$$A = \begin{bmatrix} 1 & 0 \\ -1 & 1 \end{bmatrix}.$$

What is the reciprocal of this matrix? What is $A^{1{,}000}$?

5. Show that the matrix

$$A = \begin{bmatrix} 0 & -1 \\ 1 & 0 \end{bmatrix}$$

satisfies the equation $A^2 = -I$. Use this fact to calculate the thousandth power of the matrix.

$$B = \begin{bmatrix} 1 & -1 \\ 1 & 1 \end{bmatrix}.$$

(*Hint:* Use the binomial theorem.)

3-3. The Least Equation Satisfied by a Matrix

The main theorem stated in the preceding section tells us that every $n \times n$ matrix A satisfies a polynomial equation

$$A^k + c_{k-1}A^{k-1} + \cdots + c_0 I = 0 \qquad (2)$$

with numerical coefficients, the degree k of this equation being at most n. Of course, it may satisfy some other equation of the same form also. For instance, the 4×4 matrix

$$A = \begin{bmatrix} 0 & 0 & 0 & 0 \\ 0 & 0 & 0 & 0 \\ 0 & 0 & 0 & 0 \\ 1 & 0 & 0 & 0 \end{bmatrix}$$

satisfies all the equations $A^2 = 0$, $A^3 = 0$, $A^4 = 0$. It is clear that the equation which *tells us the most* about a matrix A is the equation of *lowest degree* which A satisfies.

From among all the polynomial equations of the form (2) with numerical coefficients which a matrix A satisfies, we may always choose an equation of the smallest possible degree. We shall show in a subsequent section that there is *only one* equation

$$A^k + c_{k-1}A^k + \cdot \cdot \cdot + c_0I = 0$$

of this smallest possible degree which A satisfies. Hence, instead of speaking of *an* equation of smallest possible degree, we may speak of *the* equation of smallest possible degree, and make the following definition.

DEFINITION

The polynomial equation of smallest possible degree

$$A^k + c_{k-1}A^{k-1} + \cdot \cdot \cdot + c_0I = 0$$

with numerical coefficients $c_{k-1} \cdot \cdot \cdot c_0$ satisfied by a matrix A will be called the *least equation satisfied by* A.

It is important to realize that the polynomial equation of smallest possible degree satisfied by a particular matrix A is *unique.* When we speak of *the* least equation satisfied by A, we are *not* using the word "the" loosely: there is only one such equation. In the present section, we shall describe a systematic procedure for finding this equation.

Let us begin with 2×2 matrices. If A is such a matrix, first calculate A^2 and then start with the three matrices A^2, A, I. It is clear that, by subtracting a suitable numerical multiple of I from each of the matrices A^2 and A, we can obtain two matrices $A^2 - aI$ and $A^2 - bI$ which have the entry 0 in first place in the first row. If $A - bI = 0$, then this is the least equation satisfied by A. If $A - bI$ is different from zero, then some entry in $A - bI$, say the entry in the ith row and jth column, is different from zero. It is then clear that, by subtracting a suitable numerical multiple of $A - bI$ from the matrix $A^2 - aI$, we get a matrix $(A^2 - aI) - c(A - bI)$ which has the entry zero in the ith row and jth column.

It is then possible to prove that the matrix

$$(A^2 - aI) - c(A - bI)$$

is actually zero and that the equation

$$(A^2 - aI) - c(A - bI) = 0$$

is the least equation satisfied by A.

Let us see how this procedure works, by applying it to the matrix

$$A = \begin{bmatrix} -2 & -4 \\ 3 & 6 \end{bmatrix}.$$

If we calculate A^2, we find that

$$A^2 = \begin{bmatrix} -8 & -16 \\ 12 & 24 \end{bmatrix}.$$

Next we subtract $-2I$ from A and $-8I$ from A^2 to get the two matrices

$$A + 2I = \begin{bmatrix} 0 & -4 \\ 3 & 8 \end{bmatrix}$$

and

$$A^2 + 8I = \begin{bmatrix} 0 & -16 \\ 12 & 32 \end{bmatrix},$$

each of which has the entry zero in the first place in the first row. The matrix $A + 2I$ is not zero; so we go on. We can get a zero in the first place of the second row by subtracting $4(A + 2I)$ from $A^2 + 8I$. If we do this, we find that

$$(A^2 + 8I) - 4(A + 2I) = 0.$$

Simplifying, we find that the least equation satisfied by A is $A^2 - 4A = 0$. (Consequently, the matrix A has no reciprocal.)

Next let us consider a 3×3 matrix. If A is such a matrix, we first calculate A^2 and A^3. Then we begin with the four matrices A^3, A^2, A, I. It is clear that, by subtracting a suitable numerical multiple of I from each of the matrices A^3, A^2, and A, we can obtain three matrices $A^3 - aI$, $A^2 - bI$, and $A - cI$, all of which have the entry 0 in the first place in the first row. If $A - cI = 0$, then $A - cI = 0$ is the least equation satisfied by A. If $A - cI$ is different from zero, then some entry in $A - cI$, say the entry in the ith row and jth column, is different from zero. It is then clear that, by subtracting a suitable numerical multiple of $A - cI$ from each of the two matrices $A^3 - aI$ and $A^2 - bI$, we can get two matrices

$$(A^3 - aI) - d(A - cI) \qquad \text{and} \qquad (A^2 - bI) - c(A - cI)$$

each of which has the entry zero in the ith row and jth column. Call the second of these two matrices B. If B is zero, then

$$(A^2 - bI) - e(A - cI) = 0,$$

i.e., $B = 0$ is the least equation satisfied by A. If $B \neq 0$, then some entry in B, say the entry in the kth row and lth column, is different from zero. It is then clear that, by subtracting a suitable numerical multiple of B from the matrix $(A^3 - aI) - d(A - cI)$, we get a matrix which has the entry zero in the kth row and lth column. It is then possible to prove that this matrix is actually zero, i.e.,

$$(A^3 - aI) - d(A - cI) - f[(A^2 - bI) - e(A - cI)] = 0,$$

and that this last equation is the least equation satisfied by A.

To get some more concrete idea of how this process works, let us apply it to the 3×3 matrix

$$A = \begin{bmatrix} 1 & -1 & 0 \\ 0 & 1 & -1 \\ 1 & 0 & 1 \end{bmatrix}.$$

We first calculate A^2 and A^3:

$$A^2 = \begin{bmatrix} 1 & -2 & 1 \\ 1 & 1 & 2 \\ 2 & -1 & 1 \end{bmatrix}$$

$$A^3 = \begin{bmatrix} 2 & -3 & 3 \\ 3 & 2 & 3 \\ 3 & -3 & 2 \end{bmatrix}.$$

Now we begin to "process" the four matrices A^3, A^2, A, I. Subtracting multiples of I from each of A^3, A^2, A, we find the three matrices

$$A - I = \begin{bmatrix} 0 & -1 & 0 \\ 0 & 0 & -1 \\ 1 & 0 & 0 \end{bmatrix}$$

$$A^2 - I = \begin{bmatrix} 0 & -2 & 1 \\ -1 & 0 & -2 \\ 2 & -1 & 0 \end{bmatrix}$$

$$A^3 - 2I = \begin{bmatrix} 0 & -3 & 3 \\ -3 & 0 & -3 \\ 3 & -3 & 0 \end{bmatrix}.$$

Now subtract twice the first matrix from the second and three times the first matrix from the third so as to get two matrices each having the entry zero in the second place in the first row:

$$A^2 - 2A + I = \begin{bmatrix} 0 & 0 & 1 \\ -1 & 0 & 0 \\ 0 & -1 & 0 \end{bmatrix}$$

$$A^3 - 3A + I = \begin{bmatrix} 0 & 0 & 3 \\ -3 & 0 & 0 \\ 0 & -3 & 0 \end{bmatrix}.$$

If we subtract three times the first of these matrices from the second to get a zero entry in the third place of the first row, we find that

$$A^3 - 3A^2 + 3A - 2I = 0.$$

This shows, among other things, that A has a reciprocal and that the reciprocal (easily computed) is

$$A^{-1} = \frac{1}{2}\begin{bmatrix} 1 & 1 & 1 \\ -1 & 1 & 1 \\ -1 & 1 & 1 \end{bmatrix}.$$

The sort of procedure that we have outlined works for matrices of any size. If A is an $n \times n$ matrix, we first calculate A^2, A^3, . . . , A^n. Then we begin with the matrices A^n, A^{n-1}, . . . , A, I. It is clear that, by subtracting a suitable numerical multiple of I from each of the matrices A^n, A^{n-1}, . . . , A, we can obtain matrices $B_n = A^n - a_nI$, $B_{n-1} = A^{n-1} - a_{n-1}I$, . . . , $B_1 = A - a_1I$ all of which have the entry zero in the first place in the first row. If $B_1 = 0$, then the equation $B_1 = 0$, i.e., the equation $A - a_1I = 0$, is the least equation satisfied by A. If $B_1 \neq 0$, then some entry in B_1, say the entry in the ith row and jth column, is different from zero. It is then clear that, by subtracting a suitable numerical multiple of B_1 from each of the matrices B_n, . . . , B_2, we can obtain $C_n = B_n - b_nB_1$, $C_{n-1} = B_{n-1} - b_{n-1}B_1$, . . . , $C_2 = B_2 - b_2B_1$, each of which has the entry zero in the ith row and jth column. If $C_2 = 0$, then the equation $C_2 = 0$, i.e., the equation $(A^2 - a_2I) - b_2(A - a_1I) = 0$, is the least equation satisfied by A. If $C_2 \neq 0$, then some entry in C_2, say the entry in the kth row and lth column, is different from zero; then we subtract a suitable numerical multiple of C_2 from each of C_n, . . . , C_3, getting matrices D_n, . . . , D_3, each of which has the entry zero in the kth row and lth column. If $D_3 = 0$, then the equation $D_3 = 0$ may be proved to be the least equation satisfied by A. If $D_3 \neq 0$, we continue, getting matrices E_n, . . . , E_4, etc. It may be proved that *this process always ter-*

minates after at most n steps, and that *if Q_k is the first of the matrices B_1, C_2, D_3, E_4,* etc., *which is equal to zero, then the equation $Q_k = 0$ is the least equation satisfied by A.*

The proof, which although simple in principle is somewhat messy in detail, will be omitted.

EXERCISES

1. Find the least equation satisfied by the matrix

$$A = \begin{bmatrix} 0 & \frac{1}{2} \\ \frac{1}{3} & \frac{1}{4} \end{bmatrix}.$$

What is the reciprocal of this matrix?

2. Find the least equation satisfied by the matrix

$$A = \begin{bmatrix} 1 - \sqrt{2} & 0 \\ -1 & 3 \end{bmatrix}.$$

What is the reciprocal of this matrix?

3. Find the least equation satisfied by the matrix

$$A = \begin{bmatrix} 1 & 2 & 3 \\ 1 & -2 & 0 \\ 0 & 1 & -2 \end{bmatrix}.$$

4. Find the least equation satisfied by the matrix

$$A = \begin{bmatrix} 1 & 2 & 0 \\ 3 & 1 & 0 \\ 0 & 0 & 4 \end{bmatrix}.$$

5. Find the least equation satisfied by the matrix

$$A = \begin{bmatrix} 1 & -1 & 1 & -1 \\ 0 & 1 & 1 & 1 \\ 1 & 1 & 0 & 0 \\ 1 & -1 & 0 & 0 \end{bmatrix}.$$

3-4. Using the Least Equation Satisfied by a Matrix to Solve the Problem of Reciprocals

We saw in the preceding section how to calculate the least equation satisfied by a matrix. Once we know the least equation satisfied by a matrix A, we can easily see whether or not A has a reciprocal.

If the final coefficient c_0 in the least equation satisfied by A is different from zero, then A has a reciprocal.

PROOF

We may write the least equation satisfied by A in the form

$$A^k + c_{k-1}A^{k-1} + \cdots + c_1A = -c_0I$$

or

$$(A^{k-1} + c_{k-1}A^{k-2} + \cdots + c_1I)A = -c_0I.$$

Since the numerical coefficient c_0 is different from zero, we can divide both sides of the equation by the number $-c_0$ and find that

$$\left[-\frac{1}{c_0}(A^{k-1} + c_{k-1}A^{k-2} + \cdots + c_1I)\right]A = I.$$

This makes it plain that the matrix

$$\left[-\frac{1}{c_0}(A^{k-1} + c_{k-1}A^{k-2} + \cdots + c_1I)\right]$$

is a reciprocal of A. Q.E.D.

If the coefficient c_0 in the least equation satisfied by A is equal to zero, then A does not have a reciprocal.

PROOF

Since c_0 is zero, the least equation satisfied by A is

$$A^k + c_{k-1}A^{k-1} + \cdots + c_1A = 0.$$

This may be written as

$$A(A^{k-1} + c_{k-1}A^{k-2} + \cdots + c_1I) = 0.$$

If A had a reciprocal A^{-1} satisfying $A^{-1}A = I$, then we could multiply both sides of the above equation on the left by A^{-1} and find that

$$(A^{-1}A)(A^{k-1} + c_{k-1}A^{k-2} + \cdots + c_1 I) = 0.$$

Since $A^{-1}A = I$, this means that

$$A^{k-1} + c_{k-1}A^{k-2} + \cdots + c_1 I = 0.$$

But then A would satisfy an equation of lower degree than the least equation which is satisfied. This is impossible. Hence our assumption that A has a reciprocal matrix must be wrong. Thus A has no reciprocal matrix. Q.E.D.

We may summarize what has just been proved in the following theorem.

THEOREM

Let A be a matrix, and let

$$A^k + c_{k-1}A^{k-1} + \cdots + c_0 I = 0$$

be the least equation satisfied by A. Then A has a reciprocal if and only if the coefficient c_0 is different from zero, and if $c_0 \neq 0$, the expression

$$-\frac{1}{c_0}(A^{k-1} + c_{k-1}A^{k-2} + \cdots + c_1 I)$$

gives a reciprocal of A.

The following theorem tells us that if A has a reciprocal, then its reciprocal is unique.

***THEOREM**

The equation $AX = I$ has a solution if and only if the equation $XA = I$ has a solution. If either of these equations has a solution, then both equations have exactly one solution, and the solution of both equations is the same.

This theorem will be proved at the end of the present section.

Before you study the proof, it is well to become familiar with the significance of the theorem. For this reason, we first give some interesting consequences of the theorem, and give its proof only afterward.

The theorem that has just been stated allows us to make the following definition.

DEFINITION

If the equation $XA = I$ has a solution, then its unique solution will be called the *reciprocal* of A and written A^{-1}. If the equation $XA = I$ has no solution, then the equation $AX = I$ has no solution either. In this case we will say that A *has no reciprocal.*

It is important to realize that when we speak of "the reciprocal of A," or write the symbol A^{-1}, we are not using the word "the" loosely: there is only one reciprocal.

We may restate what has gone before in the following two theorems.

THEOREM

If A has a reciprocal A^{-1}, then

$$A^{-1}A = AA^{-1} = I.$$

THEOREM

Let A be a matrix, and let

$$A^k + c_{k-1}A^{k-1} + \cdots + c_0 I = 0$$

be the least equation satisfied by A. Then A has a reciprocal A^{-1} if and only if $c_0 \neq 0$, and in this case

$$A^{-1} = -\frac{1}{c_0}(A^{k-1} + c_{k-1}A + \cdots + c_1 I).$$

The three theorems which we have just stated provide a complete solution of the problem of reciprocals. Summarizing, we may describe the situation as follows. We do not have to distinguish between "right-hand reciprocal" and "left-hand reciprocal" (as the failure of the commutative law of multiplication might have made us fear). If A has a reciprocal, then it is related to its reciprocal through the very ordinary equations $AA^{-1} = A^{-1}A = I$. With reference to reciprocals, the main difference between matrix algebra and the ordinary algebra of real numbers is this: a real number x has a reciprocal whenever it is different from zero; a matrix A may fail to have a reciprocal even though it is different from zero. A criterion for telling just when a matrix A has a reciprocal A^{-1}, and a formula for the reciprocal in case it exists, is given by the second theorem above.

Now we shall give a proof of the starred theorem stated above. For convenience in giving the proof, we first restate the theorem in the following more detailed way.

THEOREM

Let A be a matrix, and let A have a reciprocal matrix A^{-1} satisfying $A^{-1}A = I$. Then the equation $XA = I$ has no other solution than $X = A^{-1}$. Moreover, the reciprocal matrix A^{-1} satisfies $AA^{-1} = I$ and is the only solution of the equation $AX = I$. Finally, if A has a right-hand reciprocal matrix B satisfying $AB = I$, then A also has a left-hand reciprocal A^{-1} satisfying $A^{-1}A = I$, and in fact, $A^{-1} = B$.

The proof is as follows.

PROOF

By the preceding theorem, if A has a reciprocal matrix, then the coefficient c_0 in the least equation satisfied by A is different from zero. Hence this equation can be written as

$$A^k + c_{k-1}A^{k-1} + \cdots + c_1A = -c_0I.$$

Using the law of exponents and dividing, it follows that

$$A\left[-\frac{1}{c_0}(A^{k-1} + c_{k-1}A^{k-2} + \cdots + c_1I)\right] = I.$$

Hence, if we put

$$C = -\frac{1}{c_0}(A^{k-1} + \cdots + c_1 I),$$

we find that $AC = I$. If X satisfies the equation $XA = I$, it follows, on multiplying both equations on the right by the matrix C, that

$$(XA)C = IC = C.$$

Thus

$$X(AC) = C,$$

and since $AC = I$, it follows that

$$X = XI = C.$$

Thus $X = C$ is the *only* solution of the equation $XA = I$; i.e., the equation $XA = I$ has *only one* solution. Since A^{-1} is *a solution* of the equation $XA = I$, it follows that A^{-1} is the only solution of this equation. This proves the first assertion of our theorem.

We have shown, moreover, that if A has a reciprocal matrix A^{-1}, there exists a matrix C such that $AC = I$. But then

$$A^{-1} = A^{-1}I = A^{-1}(AC) = (A^{-1}A)C = IC = C.$$

Consequently, $C = A^{-1}$, so that $AA^{-1} = I$. If $AX = I$, then $A^{-1}(AX) = A^{-1}I = A^{-1}$, so that $A^{-1} = (A^{-1}A)X = IX = X$; i.e., $X = A^{-1}$. This proves the second assertion of our theorem.

Finally, suppose that the matrix A has a right-hand reciprocal matrix B satisfying $AB = I$. Let

$$A^k + c_{k-1}A^{k-1} + \cdots + c_0 I = 0 \tag{3}$$

be the least equation satisfied by A. We shall show that $c_0 \neq 0$. If this is false, so that $c_0 = 0$, then equation (3) may be written as

$$A^k + c_{k-1}A^{k-1} + \cdots + c_1 A = 0$$

i.e., as

$$(A^{k-1} + c_{k-1}A^{k-2} + \cdots + c_1 I)A = 0.$$

Multiplying both sides of the above equation on the right by B, we find that

$$(A^{k-1} + c_{k-1}A^{k-2} + \cdots + c_1 I)(AB) = 0.$$

Since $AB = I$, this means that

$$A^{k-1} + c_{k-1}A^{k-2} + \cdots + c_1 I = 0.$$

But then A would satisfy an equation of lower degree than the least equation which it satisfies. This is impossible. Hence our assumption that $c_0 = 0$ must be wrong.

Hence, if the matrix A has a right-hand reciprocal matrix B satisfying $AB = I$, then $c_0 \neq 0$. Hence, by the previous theorem, A has a reciprocal matrix A^{-1} satisfying $A^{-1}A = I$.

Hence $B = IB = (A^{-1}A)B = A^{-1}(AB) = A^{-1}I = A^{-1}$. This proves the third and last part of our theorem. Q.E.D.

EXERCISES

1. Find the least equation satisfied by the matrix

$$A = \begin{bmatrix} 1 & 2 \\ -3 & -1 \end{bmatrix}.$$

Calculate the reciprocal of the matrix A.

2. Find the least equation satisfied by the matrix

$$A = \begin{bmatrix} 3 & 6 \\ -2 & -4 \end{bmatrix}.$$

Does this matrix have a reciprocal?

3. The least equation satisfied by the matrix

$$\begin{bmatrix} 1 & 1 & 0 \\ 0 & 1 & -1 \\ -1 & 0 & 1 \end{bmatrix}$$

is $A^3 - 3A^2 + 3A - 2I = 0$.

(*a*) Show by computing that A satisfies this equation.
(*b*) Calculate the reciprocal of A.
(*c*) Solve the equation

$$\begin{bmatrix} 1 & 1 & 0 \\ 0 & -1 & -1 \\ -1 & 0 & 1 \end{bmatrix} X = \begin{bmatrix} 1 & 2 & 3 \\ 4 & 5 & 6 \\ 7 & 8 & 9 \end{bmatrix}$$

for the matrix X.

In solving the next three exercises, use should be made of the solutions of the three final exercises of the preceding section of exercises.

4. Does the matrix

$$\begin{bmatrix} 1 & 2 & 3 \\ 1 & -2 & 0 \\ 0 & 1 & -2 \end{bmatrix}$$

have a reciprocal? If so, what is its reciprocal?

5. What is the reciprocal of the matrix

$$\begin{bmatrix} 1 & 2 & 0 \\ 3 & 1 & 0 \\ 0 & 0 & 4 \end{bmatrix}?$$

6. Does the matrix

$$\begin{bmatrix} 1 & -1 & 1 & -1 \\ 0 & 1 & 1 & 1 \\ 1 & 1 & 0 & 0 \\ 1 & -1 & 0 & 0 \end{bmatrix}$$

have a reciprocal? If so, what is its reciprocal?

3-5. Proof of the Uniqueness of the Least Equation Satisfied by a Matrix

To complete the work of Sections 3-3 and 3-4, we must prove the following theorem, which was used to prove the other theorems of Section 3-4.

THEOREM

Suppose that the matrix A satisfies two equations

$$A^k + c_{k-1}A^{k-1} + \cdots + c_0I = 0 \tag{4}$$

and

$$A^k + d_{k-1}A^{k-1} + \cdots + d_0I = 0 \tag{5}$$

of the same degree k, but satisfies no equation

$$A^j + c_{j-1}A^{j-1} + \cdots + c_0I = 0$$

of lower degree. Then each coefficient $c_{k-1}, c_{k-2}, \ldots, c_0$ is equal to the corresponding coefficient $d_{k-1}, \ldots, d_0$, so that the two equations are exactly the same.

PROOF

Suppose that our theorem is false. Then, not all the numbers $c_{k-1} - d_{k-1}, c_{k-2} - d_{k-2}, \ldots, c_0 - d_0$ are zero. Suppose that the first of these numbers (reading from left to right) which is not zero is $c_j - d_j$. Clearly, j is either $k - 1, k - 2, \ldots,$ or 1; for $j = 0$ would mean that $c_{k-1} - d_{k-1} = 0$, $c_{k-2} - d_{k-2} = 0$, $\ldots$, and $c_1 - d_1 = 0$, while $c_0 - d_0 \neq 0$ so that we would find on subtracting equation (5) from equation (4) and dividing by $c_0 - d_0$ that $I = 0$, which is false. That is, $j < k$, but $j > 0$. Then if we subtract equation (5) from equation (4) and divide by the nonzero number $c_j - d_j$, we find that

$$A^j + \left(\frac{c_{j-1} - d_{j-1}}{c_j - d_j}\right) A^{j-1} + \left(\frac{c_{j-2} - d_{j-2}}{c_j - d_j}\right) A^{j-2} + \cdots + \left(\frac{c_0 - d_0}{c_j - d_j}\right) = 0.$$

Thus the matrix A satisfies an equation of lower degree, contradicting our hypothesis. Q.E.D.

3-6. Two Theorems About Reciprocals

Let A have the reciprocal A^{-1}. Then the equation $A^{-1}X = I$ has the solution A. This means that A^{-1} has a reciprocal and that the reciprocal $(A^{-1})^{-1}$ of A^{-1} is just A. This proves the following theorem.

THEOREM

If A has a reciprocal, so does A^{-1}; and $(A^{-1})^{-1} = A$.

Suppose now that A and B are two matrices of the same size and that both A and B have reciprocals. Then, since

$$(AB)(B^{-1}A^{-1}) = A(BB^{-1})A^{-1} = AIA^{-1} = AA^{-1} = I,$$

the equation $ABX = I$ has the solution $B^{-1}A^{-1}$. This means that AB has a reciprocal and that this reciprocal is $B^{-1}A^{-1}$. This proves the following theorem.

THEOREM

If A and B are two matrices of the same size, and both A and B have reciprocals, then so does AB; and $(AB)^{-1} = B^{-1}A^{-1}$.

Note that this theorem tells us that the reciprocal of a product is the product of the reciprocals *taken in the reverse order.* Since the commutative law of multiplication fails for matrices, we must always be careful to take proper account of the order of factors when we write any expression involving a product.

We can apply the theorem to calculate the reciprocal of a product of many factors. Thus

$$\begin{aligned}(ABC)^{-1} = ((AB)C)^{-1} = C^{-1}(AB)^{-1} = C^{-1}(B^{-1}A^{-1})\\ = C^{-1}B^{-1}A^{-1},\end{aligned}$$

the product of the reciprocals in reverse order.

EXERCISES

1. Show that the matrix

$$A = \begin{bmatrix} 1 & 1 \\ 1 & 2 \end{bmatrix}$$

is the product BC of the matrices

$$B = \begin{bmatrix} 1 & 0 \\ 1 & 1 \end{bmatrix} \quad \text{and} \quad C = \begin{bmatrix} 1 & 1 \\ 0 & 1 \end{bmatrix}.$$

Use this fact to calculate A^{-1}.

2. Show that the matrix

$$D = \begin{bmatrix} 2 & 1 \\ 1 & 1 \end{bmatrix}$$

is the product CB of the matrices B and C of Exercise 1. Use this fact to solve the equation

$$X \begin{bmatrix} 2 & 1 \\ 1 & 1 \end{bmatrix} = \begin{bmatrix} 1 & 2 \\ 3 & 4 \end{bmatrix}.$$

3. Let A and B be any $n \times n$ matrices, let x be any number, and let C be any $n \times n$ matrix having a reciprocal. Show that

(*a*) $C(A + B)C^{-1} = CAC^{-1} + CBC^{-1}$
(*b*) $C(AB)C^{-1} = (CAC^{-1})(CBC^{-1})$
(*c*) $C(xA)C^{-1} = x(CAC^{-1})$
(*d*) $CIC^{-1} = I$.

4. (Continuation of 3.) Let A be any $n \times n$ matrix, and let C be any $n \times n$ matrix having a reciprocal. Show that for any positive integer j

$$CA^jC^{-1} = (CAC^{-1})^j.$$

5. (Continuation of 4.) Let A be any $n \times n$ matrix, and let C be any $n \times n$ matrix having a reciprocal. Show that if A satisfies a polynomial equation

$$X^k + d_{k-1}X^{k-1} + \cdots + d_0I = 0,$$

the coefficients $d_{k-1}, \ldots, d_0$ being numbers, then CAC^{-1} satisfies the same equation. What does this make you suspect about the number of solutions of such an equation? Can you prove it?

Chapter 4

VECTORS AND LINEAR EQUATIONS

4-1. Definition of Vectors. Notation and Properties

A certain very simple, very special kind of matrix is important in many applications. This kind of matrix is described in the following definition.

DEFINITION

An $n \times n$ matrix which has zero everywhere in every column except possibly its first column will be called a *vector* of size n. That is, a vector is a matrix which has zero everywhere in its second to last columns. Thus a matrix V of size n is a vector whenever

$$[V]_{i,2} = [V]_{i,3} = \cdot \cdot \cdot = [V]_{i,n} = 0$$

for each value of i from $i = 1$ to $i = n$.

The matrix

$$\begin{bmatrix} 1 & 0 \\ 2 & 0 \end{bmatrix}$$

is a vector of size 2; the matrix

$$\begin{bmatrix} 1 & 0 & 0 \\ 2 & 0 & 0 \\ 2 & 0 & 0 \end{bmatrix}$$

is a vector of size 3; the matrix

$$\begin{bmatrix} 1 & 0 & 0 & 0 \\ 2 & 0 & 0 & 0 \\ 2 & 0 & 0 & 0 \\ 1 & 0 & 0 & 0 \end{bmatrix}$$

is a vector of size 4; etc.

The following theorem tells us certain important facts about the relation of vectors to other kinds of matrices.

THEOREM

Let V_1 and V_2 be vectors, and let A be a matrix, all three being of the same size. Let x be a number. Then

(*a*) $V_1 + V_2$ is a vector.

(*b*) xV_1 is a vector.

(*c*) AV_1 is a vector.

PROOF

Since the entries of $V_1 + V_2$ are the sums of corresponding entries in V_1 and V_2, and since all the entries in the second, third, etc., columns of each of V_1 and V_2 are zero, it is obvious that all the entries in the second, third, etc., columns of $V_1 + V_2$ are zero. This proves part (*a*). The proof of part (*b*) is equally obvious. To prove (*c*), note that the entry in the ith row and jth column of the product AV_1 is

$$[AV_1]_{i,j} = [A]_{i,1}[V_1]_{1,j} + [A]_{i,2}[V_1]_{2,j} + \cdots + [A]_{i,n}[V_1]_{n,j}.$$

If $j > 1$, then, since V_1 is a vector, all the entries $[V_1]_{1,j}$, $[V_1]_{2,j}$, $\ldots$, $[V_1]_{n,j}$ in its jth column are zero. Therefore, all the terms in the above sum are zero, so that $[AV_1]_{i,j} = 0$ if $j > 1$; that is, all the entries in the second, third, etc., columns of AV_1 are zero. Hence, AV_1 is a vector. This proves (*c*). Q.E.D.

The following theorem gives the basic algebraic laws for calculating with matrices and vectors.

THEOREM

Let V_1 and V_2 be vectors, and let A_1 and A_2 be matrices, all four being of the same size. Let x and y be numbers. Then all the following laws are valid.

I. *Addition laws for vectors*
- I*a*. $V_1 + V_2 = V_2 + V_1$
- I*b*. $(V_1 + V_2) + V_3 = V_1 + (V_2 + V_3)$
- I*c*. $V_1 + 0 = V_1$
- I*d*. $V_1 + (-V_1) = 0.$

II. *Laws for numerical multiplication of vectors*
- II*a*. $x(V_1 + V_2) = xV_1 + xV_2$
- II*b*. $x(yV_1) = (xy)V_1$
- II*c*. $(x + y)V_1 = xV_1 + yV_1$
- II*d*. $0 \cdot V_1 = 0$
- II*e*. $1 \cdot V_1 = V_1$
- II*f*. $x \cdot 0 = 0.$

III. *Laws for multiplication of vectors by matrices*
- III*a*. $A_1(V_1 + V_2) = A_1V_1 + A_1V_2$
- III*b*. $(A_1 + A_2)V_1 = A_1V_1 + A_2V_1$
- III*c*. $A_1(A_2V_1) = (A_1A_2)V_1$
- III*d*. $0 \cdot V_1 = 0$
- III*e*. $I \cdot V_1 = V_1$
- III*f*. $A_1(xV_1) = (xA_1)V_1 = x(A_1V_1).$

PROOF

We have already proved all these laws for matrices. Since vectors are merely particular simple types of matrices, we know them to be true for vectors also. Q.E.D.

Why then is this theorem written down? Simply for review and for convenience of reference.

The entries $[V]_{1,1}, [V]_{2,1}, \ldots, [V]_{n,1}$ of a vector V of size n are more often called the components of the vector V. Since all the other entries of V are zero, it is clear that V is determined by these n components. Since $[V]_{i,j} = 0$ unless $j = 1$, it is really superfluous in dealing with vectors to write $[V]_{i,1}$ rather than simply $[V]_i$. Hence, in dealing with vectors V, we reserve the right to denote the components of V simply as $[V]_i$.

Let the components of a vector V of size n be $[V]_1 = v_1$, $[V]_2 = v_2$, etc. Then the numbers $v_1, \ldots, v_n$ determine the vector V, and it is therefore possible (also customary and often convenient) to denote the vector V by the symbol $[v_1, \ldots ,v_n]$. Thus, when we write

$$V = [v_1, \ldots ,v_n] \tag{1}$$

displaying the components of V explicitly, we really mean

$$V = \begin{bmatrix} v_1 & 0 & 0 & \cdots & 0 \\ v_2 & 0 & 0 & \cdots & 0 \\ \cdot & \cdot & \cdot & \cdots & \cdot \\ v_n & 0 & 0 & \cdots & 0 \end{bmatrix}. \tag{2}$$

But it is simpler for the printer if we write (1) rather than (2) whenever possible.

If we have two vectors

$$V = [v_1, \ldots ,v_n] \qquad \text{and} \qquad U = [u_1, \ldots ,u_n]$$

of the same size, then it is clear (since to add matrices we add corresponding entries) that

$$V + U = [v_1 + u_1, \ldots , v_n + u_n].$$

If x is a number, it is clear for the same reason that

$$xV = [xv_1, \ldots ,xv_n].$$

All this is merely old material repeated, given new names, and written a little differently. In short, nothing much. But, be patient.

4-2. Vectors and Directed Segments in the Plane

An ordered pair P, Q of points in the plane determines, geometrically, an *ordered or directed segment*, simply the segment from

P to Q. We can readily represent this in a drawing, putting an arrowhead on the Q end of the segment to show the order, thus:

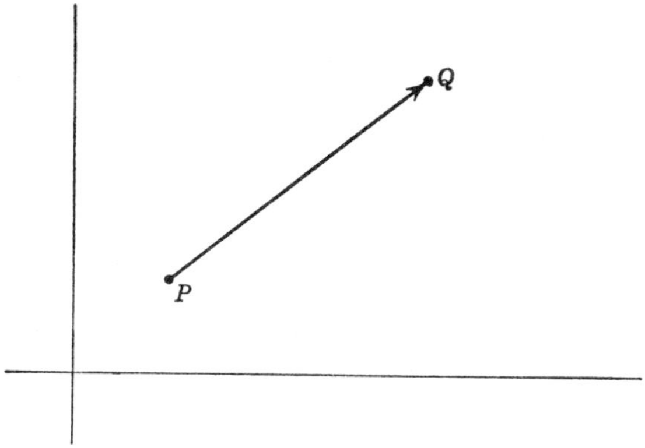

FIG. 1. The directed segment $\overrightarrow{PQ}$.

We shall write the ordered pair itself not as P, Q but as $\overrightarrow{PQ}$; this has the advantage of showing the order of the ordered pair more dramatically. Now, we know from analytic geometry that every point P in the plane has two numerical coordinates P_1 and P_2 relative to a given coordinate system. We shall use this fact to associate a vector of size 2 with every directed segment $\overrightarrow{PQ}$. The way that this is to be done is specified in the following definition.

DEFINITION

Let $\overrightarrow{PQ}$ be a directed line segment in the plane, and let P_1, P_2 and Q_1, Q_2 be the coordinates of P and of Q, respectively. Then $\overrightarrow{PQ}$ and the vector $V = [Q_1 - P_1, Q_2 - P_2]$ (evidently of size 2) are said to be associated, and $\overrightarrow{PQ}$ is said to be a geometric *representation of* V, or *to represent* V.

The vector V depends only on the differences of the coordinates of Q and P. Hence, it is clear that, if the directed segment $\overrightarrow{PQ}$

is shifted in the plane without change of direction or of length, we obtain a new directed segment $\overrightarrow{P'Q'}$ which represents the same vector as $\overrightarrow{PQ}$. Thus, a given vector V is represented not by a single directed segment, but by any directed segment which has the proper length and direction (see Fig. 2).

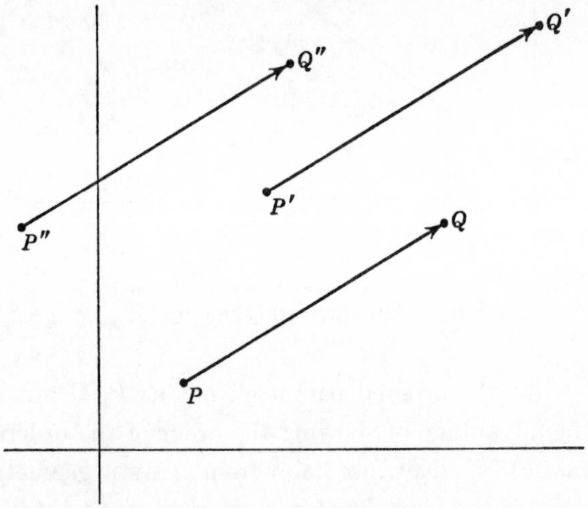

FIG. 2. Directed segments representing a given vector.

The association between directed segments and vectors which we have established permits us to give a geometric representation for the sum and difference of two vectors. This is done in the next theorem.

THEOREM ON GEOMETRIC REPRESENTATION

(*a*) If the vector V_1 is represented by the directed segment $\overrightarrow{PQ}$ and the vector V_2 is represented by the directed segment $\overrightarrow{QR}$, then the sum $V_1 + V_2$ is represented by the directed segment $\overrightarrow{PR}$ (cf. Fig. 3).

(*b*) If the vector V_1 is represented by the directed segment $\overrightarrow{PQ}$ and the vector V_2 is represented by the directed segment $\overrightarrow{PR}$, then the vector $V_1 - V_2$ is represented by the directed segment $\overrightarrow{RQ}$ (cf. Fig. 4).

(*c*) If the vector V_1 is represented by the directed segment $\overrightarrow{PQ}$, then the vector $-V_1$ is represented by the directed segment $\overrightarrow{QP}$.

Parts (*a*) and (*b*) of this theorem are illustrated by Figs. 3 and 4.

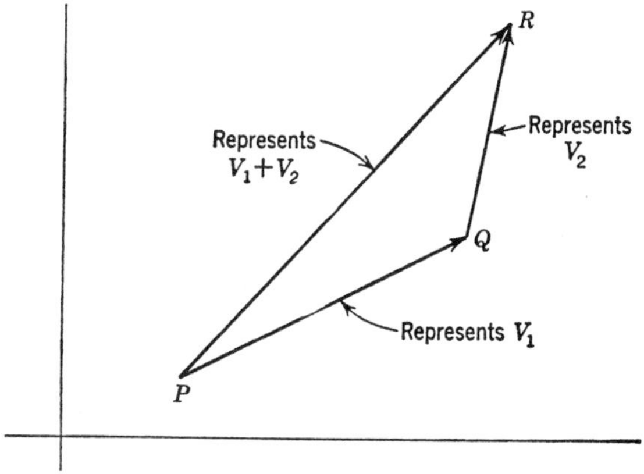

FIG. 3. The sum of two vectors.

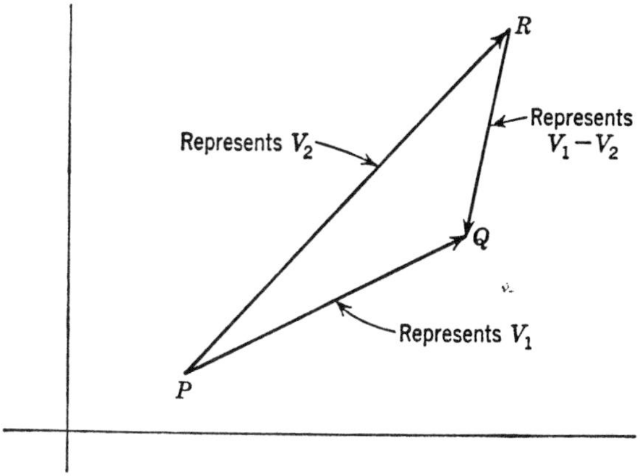

FIG. 4. The difference of two vectors.

PROOF OF THEOREM

Let P have the coordinates P_1, P_2, and Q have the coordinates Q_1, Q_2, and R have the coordinates R_1, R_2. Then $\overrightarrow{PQ}$ represents the vector $V_1 = [Q_1 - P_1,\ Q_2 - P_2]$, and $\overrightarrow{QR}$ represents the vector $V_2 = [R_1 - Q_1,\ R_2 - Q_2]$, while $\overrightarrow{PR}$ represents the vector $V_3 = [R_1 - P_1,\ R_2 - P_2]$. It is obvious on adding V_1 and V_2 that $V_1 + V_2 = V_3$. Thus $\overrightarrow{PR}$ represents $V_1 + V_2$, and this proves (a).

The proofs of (b) and (c) are equally easy, and will not be given, but will be left to the student as an exercise. Q.E.D.

The commutative law of $V_1 + V_2 = V_2 + V_1$ of addition may be given the following interesting geometric interpretation. Let

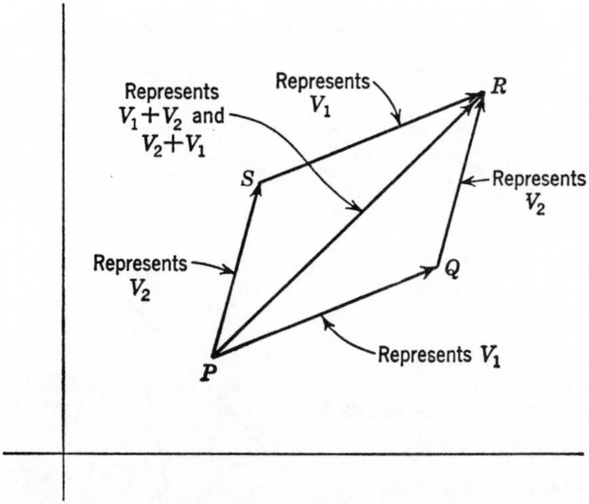

FIG. 5. The commutative law of addition of vectors.

$\overrightarrow{PS}$ be a directed segment having the same length and direction as the segment $\overrightarrow{QR}$ in Fig. 3. (See Fig. 5.) Then, by the elementary theory of parallel lines, $\overrightarrow{SR}$ has the same length and direction as $\overrightarrow{PQ}$. Since two directed segments with the same length and direction both represent the same vector, it follows that $\overrightarrow{PS}$ represents V_2 and that $\overrightarrow{SR}$ represents V_1. Thus $\overrightarrow{PR}$ represents both $V_1 + V_2$ and $V_2 + V_1$, which serves to remind us that $V_2 + V_1 = V_1 + V_2$.

Since Fig. 5 shows a parallelogram and since the diagonal of the parallelogram represents the sum $V_1 + V_2$, addition of vectors is sometimes called *addition according to the parallelogram law.*

Numerical multiples of a given vector may also be represented geometrically. This is done in the following theorem.

THEOREM

Let the directed segment $\overrightarrow{PQ}$ represent the vector V, and let x be a nonnegative real number. Then the vector xV is represented by a directed line segment having the same direction as $\overrightarrow{PQ}$ but x times as long.

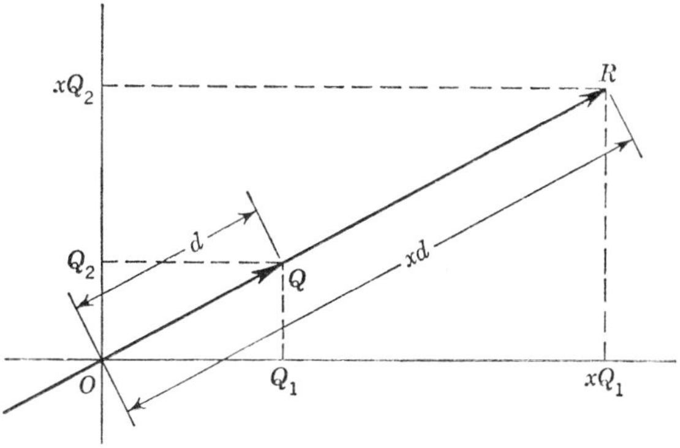

FIG. 6. A numerical multiple xV.

PROOF

Since shifting a directed line segment without change of size or direction does not change the vector V which it represents, we may suppose without loss of generality that the point P is the origin O of our coordinate system; i.e., $P = O$. Draw the infinite line OQ, and on this line in the direction OP, draw the point R for which the distance from O to R is x times the distance from O to Q, as in Fig. 6.

Then it is clear from the elementary geometry of similar triangles or, for that matter, from the basic rules of analytic geometry, that if the coordinates of the point Q are Q_1, Q_2 then the coordinates of the point R are xQ_1, xQ_2. Thus, since $\overrightarrow{OQ}$ represents V, we have $V = [Q_1,Q_2]$, and then since

$$xV = [xQ_1,xQ_2] = [xQ_1 - 0, xQ_2 - 0],$$

xV is the vector represented by the directed segment $\overrightarrow{OR}$.

Q.E.D.

To find a directed segment representing a numerical multiple xV of the vector V when x is negative, we have only to write $xV = -(-x)V$ and use both the above theorem and part (c) of the first theorem of the present section.

EXERCISES

1. Prove part (b) of the first theorem of Section 4-2.

2. Prove part (c) of the first theorem of Section 4-2.

3. Prove the following theorem: If the directed segment $\overrightarrow{PQ}$ represents the vector V, and if x is negative, then the vector xV is represented by a directed segment having $|x|$ times the length of $\overrightarrow{PQ}$ and the diametrically opposite direction.

4-3. Vectors and Directed Segments in Three-dimensional Space

The association between vectors and directed segments introduced in the preceding section is as applicable to three-dimensional space as to the two-dimensional plane. The only difference is that a directed line segment in three-dimensional space will represent a vector of size 3, not a vector of size 2. The basic definition is as follows.

DEFINITION

Let $\overrightarrow{PQ}$ be a directed line segment in three-dimensional space (i.e., an ordered pair of points in three-dimensional space), and let P_1, P_2, P_3 be the coordinates of the points P, while Q_1, Q_2, Q_3 are the coordinates of the point Q (coordinates being taken with respect to a given set of coordinate axes). Then $\overrightarrow{PQ}$ and the vector $V = [Q_1 - P_1, Q_2 - P_2, Q_3 - P_3]$ (evidently of size 3) are said to be *associated*, and $\overrightarrow{PQ}$ is said to be a *geometric representation* of V, or to *represent* V.

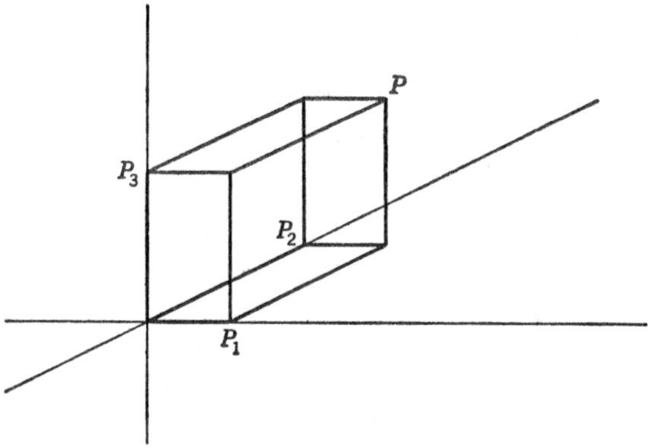

FIG. 7. The coordinates of a point in three-dimensional space.

The vector V depends only on the *differences* of the coordinates of Q and P. Thus, it is clear that, if the directed segment $\overrightarrow{PQ}$ is shifted in space without change of direction or of length, we obtain a new directed segment $\overrightarrow{P'Q'}$ which represents the same vector as $\overrightarrow{PQ}$. Thus, a given vector V is represented not by a single directed segment but by any directed segment which has the proper length and direction.

It is clear that the first theorem of the preceding section is equally true in three dimensions. Indeed, the proofs in three and in two dimensions are the same except for trivial changes.

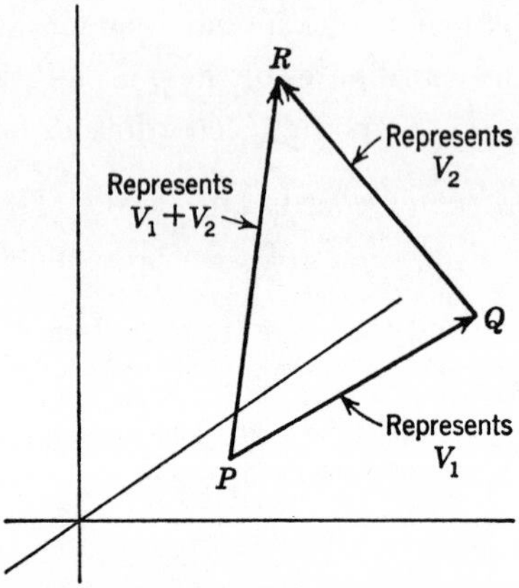

FIG. 8. The sum of two vectors.

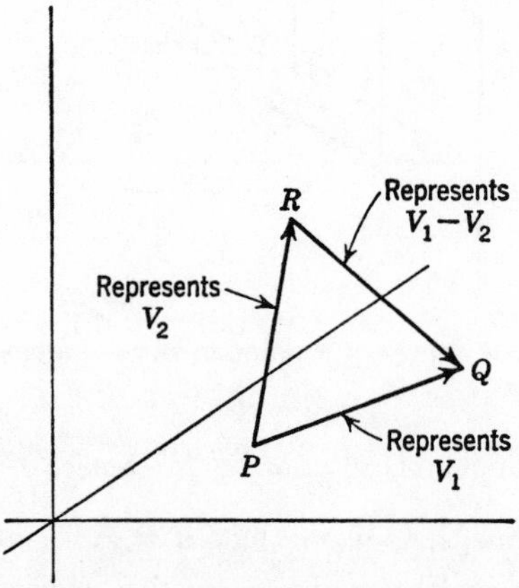

FIG. 9. The difference of two vectors.

The second theorem of the preceding section is also as true in three dimensions as in two dimensions. The proof may again be given by first supposing that the first point of the directed segment $\overrightarrow{PQ}$ representing V is the origin of coordinates, i.e.,

$P = O$, and then using the geometry of similar figures. The only complication is that we must use elementary solid geometry rather than elementary plane geometry. Rather than bothering to give a detailed proof, we shall simply refer to Fig. 10.

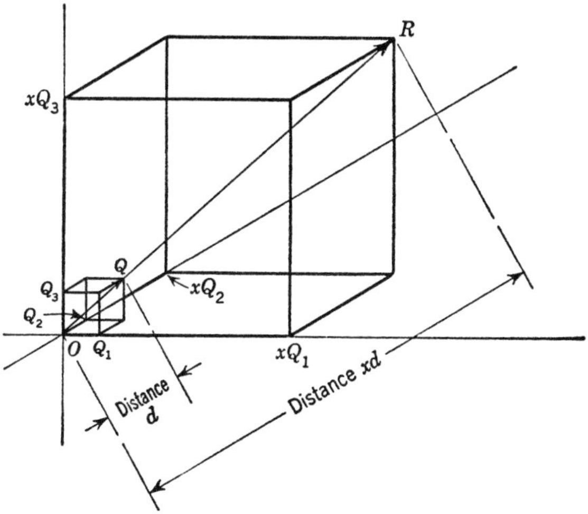

FIG. 10

To represent the vector xV when x is negative, we have again only to write $xV = -(-x)V$ and use the two theorems of Section 4-2.

EXERCISE

Prove that the first theorem of Section 4-2 remains valid for directed line segments in three-dimensional space.

4-4. Geometric Applications of the Algebra of Vectors

The connection between directed line segments and vectors explained in the two preceding theorems gives us a useful way of proving theorems in geometry by vector algebra.

Suppose that a line PQ in the plane is given and that an additional point O in the plane is also given, as in Fig. 11. Then if R is any other point on the line PQ, and if $\overrightarrow{PQ}$ represents the vector V, then by the final theorem of Section 4-2, $\overrightarrow{PR}$ represents a multiple tV of V. If $\overrightarrow{OP}$ represents a vector U, then by the first theorem of Section 4-2, $\overrightarrow{OR}$ represents the sum of the vectors represented by $\overrightarrow{OP}$ and $\overrightarrow{PR}$. That is, $\overrightarrow{OR}$ represents the vector $U + tV$. This proves the following principle.

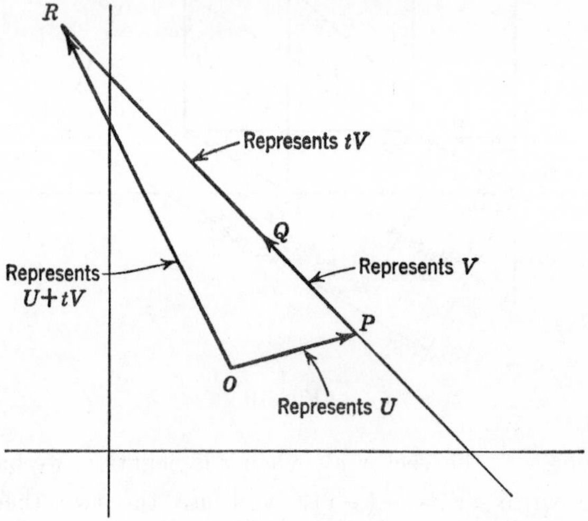

FIG. 11. Vectors to the points on a line.

PRINCIPLE A

Let l be a line in the plane and P a point on l. Let $\overrightarrow{PQ}$ be a directed line segment in l. Let O be any point in the plane. Let $\overrightarrow{PQ}$ represent the vector V and $\overrightarrow{OP}$ represent the vector U. Then each directed segment $\overrightarrow{OR}$ whose final point R lies on l represents a vector of the form $U + tV$; and each such vector is represented by exactly one directed segment $\overrightarrow{OR}$ whose final point lies on l.

(Be sure to study Fig. 11 until you fully understand the geometric meaning of this principle.)

With this principle, it is easy to prove various theorems in geometry. Let us consider as an example the following theorem.

GEOMETRIC THEOREM

The diagonals of a parallelogram bisect each other.

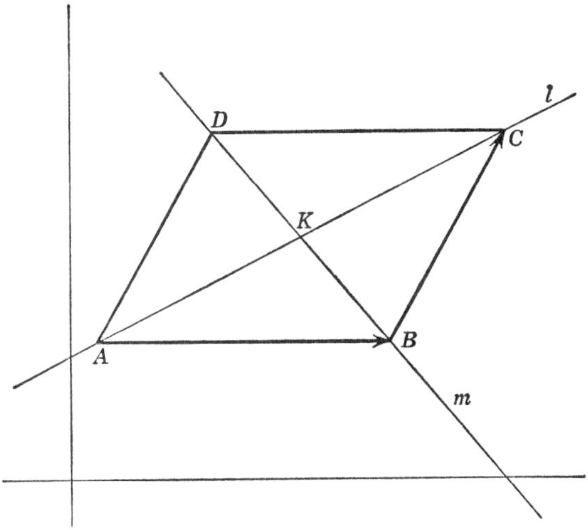

FIG. 12. The diagonals of a parallelogram bisect each other.

VECTOR-ALGEBRA PROOF

Let the parallelogram be $ABCD$. Let $\overrightarrow{AB}$ represent the vector V_1 and $\overrightarrow{BC}$ the vector V_2. Since opposite sides of a parallelogram are (parallel and) equal, $\overrightarrow{AD}$ also represents V_2 and $\overrightarrow{DC}$ represents V_1. By the first theorem of Section 4-2, $\overrightarrow{AC}$ represents $V_1 + V_2$. Similarly, $\overrightarrow{BD}$ represents $V_2 - V_1$. Let the intersection of the two diagonals be called K. Then, by the above Principle A, $\overrightarrow{AK}$ (as a point on the line l) represents a vector of the form $t(V_1 + V_2)$ and also (as a point on the line m) represents a vector of the form $V_1 + s(V_2 - V_1)$. Since $\overrightarrow{AK}$

represents both these vectors, they must be equal. Hence

$$(1 - s - t)V_1 = (t - s)V_2. \tag{3}$$

Since C does not lie on the line AB, V_1 is not a multiple of V_2. But if $t - s$ were different from zero, we could divide equation (3) by $t - s$ to find

$$\frac{(1 - s - t)}{(t - s)} V_1 = V_2,$$

so that V_2 would be a multiple of V_1. Since this is not the case, it must be that $t - s = 0$. But then

$$(1 - s - t)V_1 = 0,$$

so that $s + t = 1$. Since $t - s = 0$, $s = \frac{1}{2}$ and $t = \frac{1}{2}$. Hence $\overrightarrow{AK}$ represents $\frac{1}{2}(V_1 + V_2)$, while $\overrightarrow{AC}$ represents $V_1 + V_2$. By the theorems of Section 4-2, the length of the segment AK is exactly half that of the segment AC. Q.E.D.

The following principle is useful in giving vector-algebra proofs of geometric theorems. It could have been used to shorten the preceding proof.

PRINCIPLE B

Let AB and CD be two lines in the plane, and suppose that they are not parallel. Let $\overrightarrow{AB}$ represent the vector V_1 and $\overrightarrow{CD}$ represent the vector V_2. Then, if $sV_1 = tV_2$, both s and t are zero.

PROOF

If $a \neq 0$, then $V_1 = (t/s)V_2$, so that a directed segment $\overrightarrow{EF}$ having the *same direction* as $\overrightarrow{CD}$ (and a length equal to a/t times the length of this segment) represents the same vector as $\overrightarrow{AB}$. Thus EF is parallel to AB, and, since CD is parallel to EF, AB is parallel to CD. This contradicts our hypothesis. Hence the assumption that $s \neq 0$ is false; so $s = 0$. Similarly, $t = 0$.

Q.E.D.

The student should go back over the geometric theorem proved above and see how the principle we have just established could have been used to simplify the proof.

Once one has a little practice in giving vector-algebra proofs of geometric theorems, they are often easier to give than ordinary geometric proofs. A number of geometric theorems which we can easily prove by vector algebra are given in the following set of exercises.

EXERCISES

Prove the following geometric theorems using vector algebra:

1. The median of a trapezoid is parallel to the bases and equal to one-half their sum.

2. The three medians of a triangle are concurrent.

3. The line joining the mid-points of the two diagonals of a trapezoid is parallel to the bases and equal to one-half their difference.

4. Two medians of a triangle intersect at a point two-thirds of the distance along either median from the vertex through which it passes.

5. In parallelogram $ABCD$, AE is one-half of AB. Prove that AF is one-third of AC. (See Fig. 13.)

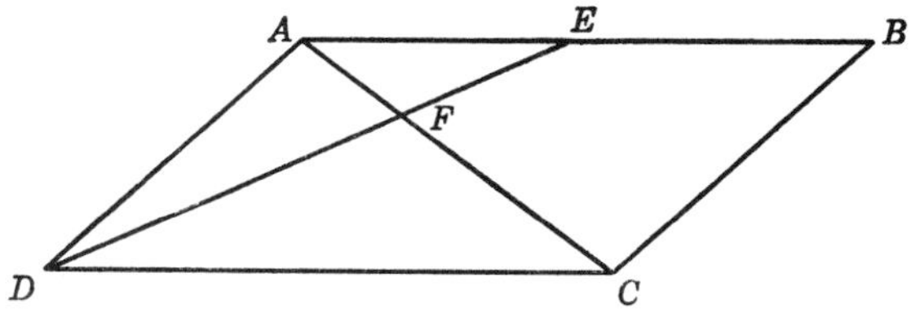

FIG. 13

6. (Ceva's theorem). In the following diagram (Fig. 14) the lines AA', BB', and CC' are concurrent if and only if the lengths $\overset{\frown}{AB'}$, $\overset{\frown}{CB'}$, $\overset{\frown}{CA'}$, $\overset{\frown}{BA'}$, $\overset{\frown}{BC'}$, and $\overset{\frown}{AC'}$ satisfy the equation

$$\overset{\frown}{AB'} \cdot \overset{\frown}{CA'} \cdot \overset{\frown}{BC'} = \overset{\frown}{AC'} \cdot \overset{\frown}{BA'} \cdot \overset{\frown}{CB'}.$$

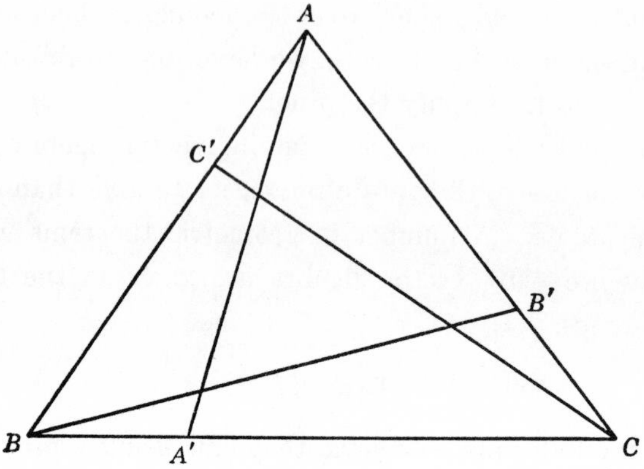

FIG. 14

4-5. Distances, Cosines, and Vectors

If the directed segment $\overrightarrow{PQ}$ in the plane represents the vector V, then $V = [Q_1 - P_1, Q_2 - P_2]$; P_1, P_2 and Q_1, Q_2 being the coordinates of the points P and Q, respectively. Hence, by the Pythagorean theorem (the distance formula for coordinate geometry) the length of $\overrightarrow{PQ}$ is the square root of the sum of the squares $(Q_1 - P_1)^2$ and $(Q_2 - P_2)^2$; that is, the length of $\overrightarrow{PQ}$ is the square root of the sum of the squares of the components of the vector V represented by $\overrightarrow{PQ}$.

We state this as a formal theorem.

THEOREM

If the directed segment $\overrightarrow{PQ}$ in the plane represents the vector $V = [v_1,v_2]$, then the length of $\overrightarrow{PQ}$ is

$$(v_1^2 + v_2^2)^{\frac{1}{2}}.$$

The trigonometric law of cosines leads to another interesting algebraic expression. Let ABC be a triangle. Let $\overrightarrow{AB}$ represent the vector V and $\overrightarrow{AC}$ represent the vector U. Then, by the law of cosines, we have the relation

$$\widehat{BC}^2 = \widehat{AC}^2 + \widehat{AB}^2 - 2\widehat{AB}\widehat{AC} \cos \measuredangle A$$

between the lengths $\widehat{BC}$, $\widehat{AC}$, and $\widehat{AB}$. Using our previous theorem we find that if $V = [v_1,v_2]$ and $U = [u_1,u_2]$, then since $\overrightarrow{BC}$ represents $U - V$, the law of cosines gives

$$(v_1 - u_1)^2 + (v_2 - u_2)^2 = (v_1^2 + v_2^2) + (u_1^2 + u_2^2) - 2\widehat{AB}\widehat{AC} \cos \measuredangle A.$$

Multiplying out and canceling redundant terms, we have

$$u_1v_1 + u_2v_2 = \widehat{AB}\widehat{AC} \cos \measuredangle A.$$

This formula proves the following theorem.

THEOREM

Let the directed segments $\overrightarrow{AB}$ and $\overrightarrow{AC}$ represent the vectors $[v_1,v_2]$ and $[u_1,u_2]$, respectively. Then the expression $v_1u_1 + v_2u_2$ gives the product of the lengths of the segments multiplied by the cosine of the angle between them.

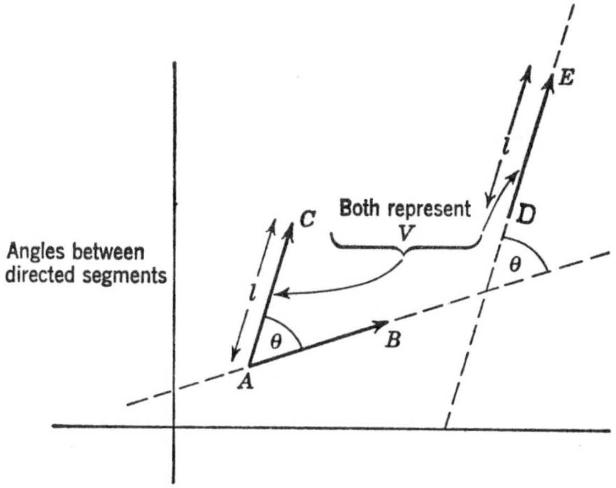

FIG. 15. Angles between directed segments.

If we shift the directed segment $\overrightarrow{AC}$ to a new position, without changing its length or direction, obtaining a new directed segment $\overrightarrow{DE}$, then it continues to represent the same vector $[v_1,v_2]$. Moreover, its length and the angle it makes with the line AC remain unchanged. Thus, in the preceding theorem, there is really no reason why $\overrightarrow{AB}$ and $\overrightarrow{AC}$ should have to have the same initial point; all that matters are the lengths and direction of these two segments (cf. Fig. 15). It follows from these considerations that the preceding theorem can be stated in the following more general, and more useful, form.

THEOREM

Let the directed segments $\overrightarrow{AB}$ and $\overrightarrow{DE}$ represent the vectors $[v_1,v_2]$ and $[u_1,u_2]$, respectively. Then the expression $v_1u_1 + v_2u_2$ gives the product of the lengths of the two segments multiplied by the cosine of the angle between them.

Two directed line segments are perpendicular if and only if the cosine of the angle between them is zero. Thus, the preceding theorem has the following theorem as an immediate consequence.

THEOREM

Let the directed segments $\overrightarrow{AB}$ and $\overrightarrow{CD}$ represent the vectors $[v_1,v_2]$ and $[u_1,u_2]$, respectively. Then $\overrightarrow{AB}$ and $\overrightarrow{CD}$ are perpendicular if and only if

$$u_1v_1 + u_2v_2 = 0.$$

Exactly corresponding formulae hold for directed segments in three-dimensional space and their corresponding vectors. To see this, we must first realize that the three-dimensional analog of the theorem of Pythagoras is:

The length of the diagonal of a rectangular parallelepiped is the square root of the sum of the square of the lengths of any set of three perpendicular sides. (Cf. Fig. 16.)

The proof is easy. By the ordinary Pythagorean theorem

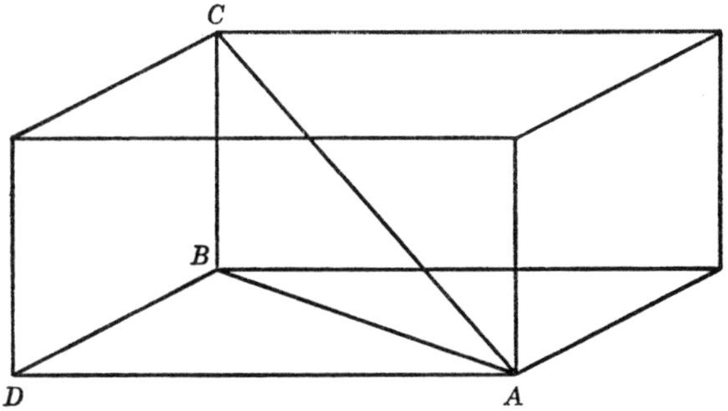

FIG. 16. The length of the diagonal.

in the plane, the distances $\widehat{AB}$, $\widehat{AD}$, and $\widehat{BD}$ are related by $\widehat{AB}^2 = \widehat{AD}^2 + \widehat{BD}^2$. Applying this same theorem to the right triangle ABC, we find that $\widehat{AC}^2 = \widehat{AB}^2 + \widehat{BD}^2$. Substituting for $\widehat{AB}^2$, we get the desired formula $\widehat{AC}^2 = \widehat{AD}^2 + \widehat{DB}^2 + \widehat{BC}^2$.

Once we have this three-dimensional form of Pythagorean theorem, it follows at once that the distance between the two points P and Q in three-dimensional space with the coordinates P_1, P_2, P_3 and Q_1, Q_2, Q_3 is the square root of $(Q_1 - P_1)^2 + (Q_2 - P_2)^2 + (Q_3 - P_3)^2$. This is clear from the following diagram (Fig. 17):

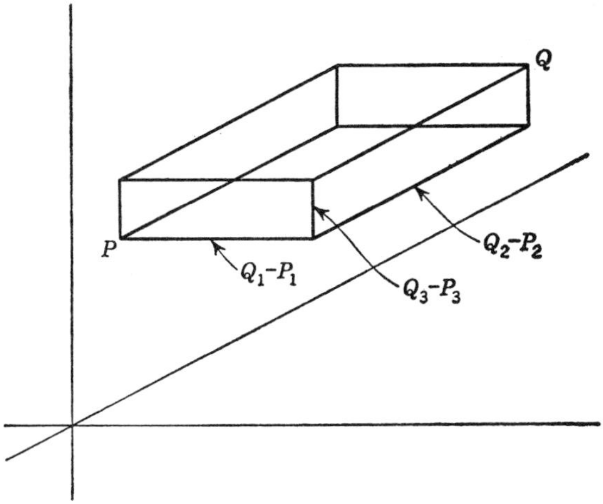

FIG. 17. Differences of the coordinates of P and Q.

Once we know this formula, we can prove the following theorems for directed segments in three-dimensional space almost exactly as they were proved for directed segments in the plane.

THEOREM

If the directed segment $\overrightarrow{PQ}$ in three-dimensional space represents the vector $V = [v_1,v_2,v_3]$, then the length of $\overrightarrow{PQ}$ is $(v_1^2 + v_2^2 + v_3^2)^{\frac{1}{2}}$.

THEOREM

Let the directed segments $\overrightarrow{AB}$ and $\overrightarrow{CD}$ represent the vectors $[v_1,v_2,v_3]$ and $[u_1,u_2,u_3]$, respectively. Then the expression $v_1u_1 + v_2u_2 + v_3u_3$ gives the product of the lengths of the two segments multiplied by the cosine of the angle between them.

THEOREM

Let the directed segments $\overrightarrow{AB}$ and $\overrightarrow{CD}$ represent the vectors $[v_1,v_2,v_3]$ and $[u_1,u_2,u_3]$, respectively. Then $\overrightarrow{AB}$ and $\overrightarrow{CD}$ are perpendicular if and only if

$$u_1v_1 + u_2v_2 + u_3v_3 = 0.$$

By virtue of the importance of expressions like $(v_1^2 + v_2^2 + v_3^2)^{\frac{1}{2}}$ and $u_1v_1 + u_2v_2 + u_3v_3$, it is natural to make the following definition, which we immediately generalize to vectors of arbitrary size.

DEFINITION

Let $U = [u_1, \ldots ,u_n]$ and $V = [v_1, \ldots ,v_n]$ be two vectors of the same size. Then

(*a*) The quantity $(v_1^2 + v_2^2 + \cdots + v_n^2)^{\frac{1}{2}}$ (the positive square root is meant) will be called the *length* of V and written $\|V\|$.

(*b*) The quantity $u_1v_1 + \cdots + u_nv_n$ will be called the *circle product* of U and V and written $U \circ V$.

It should be noted that the *circle product of two vectors is a number* (not a matrix, or a vector, or anything else).

Since the quantity $\|V\|$ is to be defined as a *positive square root,* it is quite clear that $\|V\| \geq 0$.

The two notions of length of a vector and circle product have a number of simple but highly useful algebraic properties, which we collect in the following theorem.

THEOREM

Let

$$U = [u_1, \ldots, u_n],\ V = [v_1, \ldots, v_n], \text{ and } W = [w_1, \ldots, w_n]$$

be three vectors all of the same size; and let x be a real number. Then

(*a*) $\|v\| = (V \circ V)^{\frac{1}{2}}$
(*b*) $\|xV\| = |x|\ \|V\|$
(*c*) $(xU) \circ V = U \circ (xV) = x(U \circ V)$
(*d*) $U \circ (V_1 + V_2) = U \circ V_1 + U \circ V_2$
(*e*) $U \circ V = V \circ U$
(*f*) $\|V\| \geq 0$.

Part (*c*) enables us "to take a constant out of" a circle product of two vectors. Part (*d*) enables us to "multiply out" a circle product. Part (*e*) enables us to switch the order of factors.

PROOF

(*f*) has already been noted; (*e*) simply says

$$u_1v_1 + \cdots + u_nv_n = v_1u_1 + \cdots + v_nu_n.$$

(*d*) simply says

$$u_1(v_1 + u_1) + \cdots + u_n(v_n + w_n) = (u_1v_1 + \cdots + u_nv_n) + (u_1w_1 + \cdots + u_nw_n).$$

(*c*) simply says

$$\begin{aligned}(xu_1)v_1 + \cdots + (xu_n)v_n &= u_1(xv_1) + \cdots + u_n(xv_n)\\ &= x(u_1v_1 + \cdots + v_nx_n).\end{aligned}$$

(*a*) simply says

$$(u_1^2 + \cdots + u_n^2)^{\frac{1}{2}} = (u_1u_1 + u_2u_2 + \cdots + u_nu_n)^{\frac{1}{2}}.$$

(*b*) may be proved as follows: by (*a*) and (*c*)

$$\begin{aligned}\|xU\| &= (xU \circ xU)^{\frac{1}{2}} = (x^2(U \circ U))^{\frac{1}{2}}\\ &= (x^2)^{\frac{1}{2}}(U \circ U)^{\frac{1}{2}} = |x|\ \|V\|.\end{aligned}$$

Q.E.D.

The theorems proved in this section enable us to prove a wide variety of geometric theorems by using vector algebra. As an example, we will prove the following geometric theorem.

GEOMETRIC THEOREM

The diagonals of a rhombus are perpendicular.

PROOF BY VECTOR ALGEBRA (*cf. Fig. 18*)

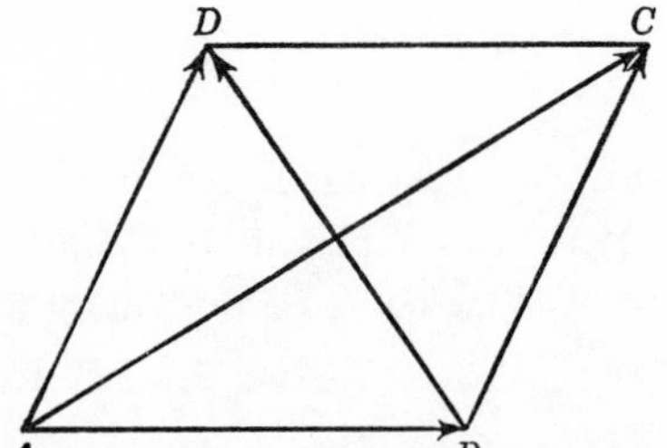

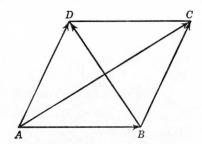

FIG. 18. The diagonals of a rhombus.

Let the rhombus be $ABCD$. Let $\overrightarrow{AB}$ represent the vector U, and let $\overrightarrow{BC}$ represent the vector V. Then $\overrightarrow{AC}$ represents $U + V$. Since $\overrightarrow{AD}$ has the same direction and length as $\overrightarrow{BC}$, it also represents the vector V. Consequently, $\overrightarrow{BD}$ represents $V - U$.

Now, $(V - U) \circ (V + U)$ multiplied out gives

$$V \circ V - U \circ U = \|V\|^2 - \|U\|^2.$$

By the first theorem of this section, this is equal to the difference of the squares of distances $AB^2 - BC^2$. Since all sides of a rhombus are equal, it follows at once that

$$(V - U) \circ (V + U) = 0.$$

Hence, by the third theorem of the present section, $\overrightarrow{AC}$ and $\overrightarrow{BD}$ are perpendicular. Q.E.D.

A number of additional geometric theorems that are easy to prove using vector algebra are given in the following set of exercises.

EXERCISES

1. Prove that for any two vectors of the same size

$$\|U - V\|^2 + \|U + V\|^2 = 2(\|U\|^2 + \|V\|^2).$$

2. Prove that if a vector U has length zero, then $U = 0$.

3. If the directed segments $\overrightarrow{AB}$ and $\overrightarrow{CD}$ in the plane are perpendicular, and if $\overrightarrow{AB}$ represents the vector $[u_1,u_2]$, then $\overrightarrow{CD}$ represents a vector which is a numerical multiple of the vector $[-u_2,u_1]$.

Prove the following geometric theorems by using vector algebra:

4. The perpendicular bisectors of a triangle are concurrent.

5. If the diagonals of a parallelogram are perpendicular, the parallelogram is a rhombus.

6. If the bisector of an angle of a triangle is perpendicular to the opposite side, the triangle is isosceles.

4-6. More about the Lengths of Vectors

Let V and U be two vectors of size 2. Let the directed segment $\overrightarrow{PQ}$ represent V, and let the directed segment $\overrightarrow{QR}$ represent U.

Then the directed segment $\overrightarrow{PR}$ represents $V + U$. Since according to the theorems of Section 4-4 the length PR is $\|V + U\|$, the length PQ is $\|V\|$, and the length QR is $\|U\|$, it follows from the geometric theorem that the length of any side of a triangle is less than or equal to the sum of the lengths of the other two sides, that $\|V + U\| \leq \|V\| + \|U\|$. The same proof works just as well for vectors of size 3. Thus, if V and U are any vectors of size 3, $\|V + U\| \leq \|V\| + \|U\|$.

If we write these inequalities out, using the detailed formulae for length given in the preceding section, we see that what we have proved are the following famous and interesting inequalities:

Let $u_1, u_2, u_3, v_1, v_2, v_3$ be any six real numbers. Then

$$(a)\ [(u_1 + v_1)^2 + (u_2 + v_2)^2]^{\frac{1}{2}} \leq (u_1^2 + u_2^2)^{\frac{1}{2}} + (v_1^2 + v_2^2)^{\frac{1}{2}}$$

$$(b)\ [(u_1 + v_1)^2 + (u_2 + v_2)^2 + (u_3 + v_3)^2]^{\frac{1}{2}} \leq (u_1^2 + u_2^2 + u_3^2)^{\frac{1}{2}} + (v_1^2 + v_2^2 + v_3^2)^{\frac{1}{2}}.$$

The fact that we are able to prove this inequality for vectors of sizes 2 and 3 suggests strongly that it is true for vectors of any size. Of course, for vectors of size greater than 3 we can no longer resort to geometry. But, as we shall now see, we can still prove the same inequality. We begin with a preliminary theorem.

THEOREM

Let U and V be vectors of the same size. Then

$$|(U \circ V)| \leq \|U\| \circ \|V\|;$$

that is, the absolute value of the circle product is at most equal to the product of the lengths.

This inequality is also very famous. It is generally known as Schwarz's inequality, after one of its discoverers, the German mathematician Hermann Amandus Schwarz.

PROOF

Let x be any number. Then, by (*a*) of the last theorem of Section 4-5,

$$(U + xV) \circ (U + xV) = (\|U + xV\|)^2 \geq 0$$

(since the square of any real number is nonnegative). Multiplying out, we find that

$$U \circ U + x^2(V \circ V) + 2x(U \circ V) \geq 0 \qquad \text{for every } x. \tag{4}$$

We will consider two cases separately:

Case 1. $V \circ V = 0$.
Case 2. $V \circ V \neq 0$.

If $V \circ V = 0$, then it follows that

$$U \circ U + 2x(U \circ V) \geq 0 \qquad \text{for every } x.$$

For such a linear expression to be nonnegative for all x's, it must be constant; otherwise its graph would surely cross the x axis at least once. Hence the expression $U \circ U + 2x(U \circ V)$ is constant, i.e., $U \circ V = 0$. In this case, the inequality

$$|(U \circ V)| \leq \|U\| \, \|V\|$$

just says

$$0 \leq \|U\| \, \|V\|,$$

and since $\|V\| = (V \circ V)^{\frac{1}{2}} = 0^{\frac{1}{2}} = 0$, this is surely true. Thus case 1 is settled affirmatively.

If $V \circ V \neq 0$, then we put

$$x = -\frac{U \circ V}{V \circ V}$$

in the inequality (4). Multiplying through both sides by the nonnegative quantity $V \circ V = \|V\|^2$, we find that

$$(U \circ U)(V \circ V) + (U \circ U)(V \circ V) - 2(U \circ V)^2 \geq 0.$$

Transposing and dividing by 2 and noting that $U \circ U = \|U\|^2$,

$V \circ V = \|V\|^2$, we have

$$(\|U\| \, \|V\|)^2 \geq (U \circ V)^2.$$

Taking the square root of both sides of this inequality, we find that

$$\|U\| \, \|V\| \geq \{(U \circ V)^2\}^{\frac{1}{2}} = |(U \circ V)|.$$

Thus case 2 is settled affirmatively also, and our theorem is proved in every possible case. Q.E.D.

Now we shall use the preceding theorem to prove the desired inequality.

THEOREM

Let U and V be vectors of the same size. Then

$$\|U + V\| \leq \|U\| + \|V\|.$$

PROOF

By (*a*) of the last theorem of Section 4-5, we have upon multiplying out

$$\begin{aligned} \|U + V\|^2 = (U + V) \circ (U + V) &= U \circ U + V \circ V + 2U \circ V \\ &= \|U\|^2 + \|V\|^2 + 2U \circ V. \end{aligned}$$

Now, any number x satisfies the inequality $x \leq |x|$; this inequality is obvious if x is positive, in which case $|x| = x$, and obvious also when x is negative, in which case it simply says that the negative quantity x is less than the positive quantity $|x|$. Consequently, if in the final sum in the last displayed equation we replace $U \circ V$ by $|U \circ V|$, we are replacing one term of a sum by a larger term and, hence, get a larger sum. Thus

$$\|U + V\|^2 \leq \|U\|^2 + \|V\|^2 + 2|U \circ V|.$$

Since by the preceding theorem $\|U \circ V\| \leq \|U\| \, \|V\|$, we may replace $|U \circ V|$ on the right by $\|U\| \, \|V\|$ and get an even larger sum. Hence

$$\|U + V\|^2 \leq \|U\|^2 + \|V\|^2 + 2\|U\| \, \|V\|.$$

That is,

$$\|U + V\|^2 \leq (\|U\| + \|V\|)^2.$$

Taking the square root of both sides of this inequality, we find that

$$\|U + V\| \leq \|U\| + \|V\|.$$

Q.E.D.

EXERCISES

1. Prove that if $|(U \circ V)| = \|U\| \, \|V\|$ and $V \neq 0$, then U is a multiple of V. (*Hint:* Consider the expression $\|U + xV\|^2$.) What does this mean geometrically?

2. Prove that if $\|U\| + \|V\| = \|U + V\|$ and $V \neq 0$, then U is a multiple of V. (*Hint:* Consider the expression $\|U + V\|^2$, and use Exercise 1.) What does this mean geometrically?

3. Prove that $\|U + V + W\| \leq \|U\| + \|V\| + \|W\|$. Can you generalize this inequality?

4-7. Expressing Systems of Linear Equations in Terms of Vectors and Matrices. Solving by Using Reciprocal Matrices

Suppose that we have a system of n linear equations for n unknowns:

$$\begin{array}{l} a_{11}u_1 + a_{12}u_2 + \cdots + a_{1n}u_n = v_2 \\ a_{21}u_1 + a_{22}u_2 + \cdots + a_{2n}u_n = v_2 \\ \cdots\cdots\cdots\cdots\cdots\cdots\cdots\cdots \\ a_{k1}u_1 + a_{k2}u_2 + \cdots + a_{2n}u_n = v_k \\ \cdots\cdots\cdots\cdots\cdots\cdots\cdots\cdots \\ a_{n1}u_1 + a_{n2}u_2 + \cdots + a_{nn}u_n = v_n. \end{array} \qquad (5)$$

This set of equations may be expressed very neatly in terms of matrices. Suppose that we let A be the matrix which has the entry a_{ij} in its ith row and jth column. (This matrix is called the coefficient matrix for the system of equations.) Let U be the vector $u_1, \ldots, u_n$. Then AU is also a vector. What are the components of this vector? To answer this question, we have simply to compute the entries in the first column of AU (since the entries in every other column are all zero). The entry in the kth row, first column of AU is, by definition of matrix multiplication,

$$a_{k1}u_1 + \cdots + a_{kn}u_n.$$

That is precisely v_k. We consequently have established the following principle:

If A is the matrix which has the entry a_{ij} in its ith *row and* jth *column, while U and V are the vectors $U = [u_1, \ldots, u_o]$ and $V = [v_1, \ldots, v_n]$, then the system of equations* (5) *may be expressed in terms of matrices as*

$$AU = V. \tag{6}$$

This principle shows us how far we have come. By introducing arrays of numbers as objects in their own right, and by defining suitable algebraic operations among those arrays, we are able to write the complicated system of equations (5) in the "elementary" form (6). Use of this artfully condensed notation gives us a general view of the system (5), as it were: we recognize the system (5) to be a grown-up cousin of the equation $au = v$ which we learned to solve "by division" in the first year of algebra. And this gives us the idea for solving the system (5): simply solve "by division." We know that this will work if we "can divide by A," that is, if A has a reciprocal. In this case, we just multiply both sides by the reciprocal of A and get the solution

$$U = A^{-1}V.$$

Our powerful machinery has done all the work for us. We have succeeded in our aim, which is perhaps the general aim of all mathematics: to make the complicated simple by discovering its inner pattern.

We summarize in the following way: *A system of n linear equations for n unknowns may be written in terms of its coefficient matrix A as the equation $AU = V$ for the unknown vector U. If A has a reciprocal matrix, then the solution is given by $U = A^{-1}V$.*

EXERCISES

Solve the following systems of linear equations by computing inverse matrices.

1.
$$\begin{aligned} x + 2y + 3z &= 0 \\ 4x + 5y + 6z &= 1 \\ 7x + 8y &= 0. \end{aligned}$$

2.
$$\begin{aligned} v + x + y - 2z &= 1 \\ v - x + y - 2z &= 0 \\ v + 2x - y + 2z &= 0 \\ v - x - y + 4z &= 1. \end{aligned}$$

(The teacher may invent and assign additional exercises of this same sort.)

4-8. Solving Systems of Linear Equations by the Method of Elimination

The method of "matrix reciprocals" for solving systems of linear equations presented in the preceding section, although it has the advantage of indicating in a very direct way the basic structure of a system of linear equations, has a number of significant disadvantages.

1. It works only when there are as many equations as unknowns.
2. It works only when the coefficient matrix has a reciprocal.
3. It is unnecessarily cumbersome from the computational point of view.

In the present section, we will develop a method for solving systems of linear equations which is free of these objections.

The method we will develop, called the *method of elimination*, is perhaps the swiftest and best method for solving these systems of equations. It is the method most commonly used in solving systems of many equations for many unknowns on ultra-high-speed computing machines of the electronic-brain type.

We begin with the following definition: Two systems of equations are called *equivalent* if every solution of one system is a solution of the other system, and vice versa.

Before taking up the most general case, let us first consider a system of four linear equations for four unknowns.

$$\begin{aligned} 2v + x + 2y - 3z &= 1 \\ 4v + x + y + z &= 2 \\ 6v - x - 5y + z &= 3 \\ 4v - 2x + 3y - z &= 0. \end{aligned}$$

We multiply the first equation by 2 and subtract from the second equation; multiply the first equation by 3 and subtract from the third equation; multiply the first equation by 2 and subtract from the fourth equation. This gives the following system of equations:

$$\begin{aligned} 2v + x + 2y - 3z &= 1 \\ 0 - x - 3y + 7z &= 0 \\ 0 - 4x - 11y + 10z &= 1 \\ 0 - 4x - y + 5z &= -2. \end{aligned}$$

We can obviously get the first system of equations back from the second just by adding twice the first equation to the second, twice the first equation to the fourth, and three times the first equation to the third. Hence the first set of equations is equivalent to the second set of equations. The second set, however, has the advantage that the *unknown v occurs* (with a coefficient different from zero) *only in the first equation.* Now we repeat this process. We add the second equation to the first, subtract four times the second equation from the third, and subtract four times the second equation from the fourth. This gives

$$\begin{aligned} 2v + 0 - y + 4z &= 1 \\ 0 - x - 3y + 7z &= 0 \\ 0 + 0 + y - 18z &= 1 \\ 0 + 0 + 11y - 23z &= -2. \end{aligned}$$

For the same reason as before, this system of equations is equivalent to the systems from which it was derived. But now each of the variables v and x occurs (with a coefficient different from zero) only in one equation.

It is clear that we can continue this process. Multiplying the third equation by suitable numbers and subtracting it from the others, we find the equivalent system

$$\begin{aligned} 2v + 0 + 0 - 14z &= 2 \\ 0 - x + 0 - 47z &= 3 \\ 0 + 0 + y - 18z &= 1 \\ 0 + 0 + 0 - 221z &= 9. \end{aligned}$$

In this system, each of the three variables v, x, y occurs (with a coefficient different from zero) in only one equation. Now we subtract suitable multiples of the fourth equation from each of the others: it is clear that in this way we will arrive at an equivalent system in which each variable occurs in only one equation. And now we can solve simply by dividing.

It is plain that this systematic elimination process can be applied to systems of any number of linear equations with any number of unknowns. Let us examine the general situation and see what can happen. Suppose that we have a system of linear equations in certain unknowns. If any unknown occurs with the coefficient zero in every equation, it plays no role and we drop it. Then we apply the subtraction process described above a certain number of times, say k times. At the end of this process, we arrive at an equivalent system of equations in which k of our unknowns occur (with a coefficient different from zero) in exactly one equation. That is, calling these unknowns $x_1, \ldots, x_k$, we arrive at an equivalent set of equations of the form

a_1x_1 + linear terms in unknowns other than $x_1, \ldots, x_k = b_1$
a_2x_2 + linear terms in unknowns other than $x_1, \ldots, x_k = b_2$
. . . .
a_kx_k + linear terms in unknowns other than $x_1, \ldots, x_k = b_k$
and "other equations" in which the unknowns $x_1, \ldots, x_k$ do not appear.

If any variable occurs with a nonzero coefficient in any one of the "other equations," we can continue our elimination process. Eventually we will come to an end. At this point, our system of equations must look like this:

a_1x_1 + linear terms in unknowns other than $x_1, \ldots, x_k = b_1$
a_2x_2 + linear terms in unknowns other than $x_1, \ldots, x_k = b_2$
. . . .
a_kx_k + linear terms in unknowns other than $x_1, \ldots, x_k = b_k$
"other equations" in which no variable appears with a nonzero coefficient.

(7)

What can one of these other equations, which must be a linear equation in which no variable appears, look like? It must be either of the form $0 = 0$, in which case we might as well drop it; or of the form $0 = b$, where b is a constant different from zero, in which case it is a contradiction. Hence we see that: *either* the "other equations" all state simply $0 = 0$, in which case they can be dropped; *or* one of the "other equations" is an obvious contradiction. Since the system of equations (7) is equivalent to the original system of equations, (7) can contain only contradictions, i.e., state impossibilities, if the original system of equations also states impossibilities, i.e., if the original system of equations simply has no solutions.

Thus, if the "other equations" in (7) are not simply $0 = 0$

repeated several times, the original system of equations has no solutions.

Summarizing, we see that we have established the following:

If the process of elimination by subtraction described above is repeatedly applied to an arbitrary system of linear equations and carried through to the end, then we will arrive at either (*a*) or (*b*):

(*a*) *A number of palpable contradictions of the form* $0 = b$, b *being some nonzero real number. In this case, the original system of equation has no solution.*

(*b*) *An equivalent system of equations of the form*

$$\begin{aligned} a_1x_1 &+ \text{linear terms in unknowns other than } x_1, \ldots, x_k = b_1 \\ a_2x_2 &+ \text{linear terms in unknowns other than } x_1, \ldots, x_k = b_2 \\ &\cdots\cdots\cdots\cdots\cdots\cdots\cdots\cdots \\ a_kx_k &+ \text{linear terms in unknowns other than } x_1, \ldots, x_k = b_k, \end{aligned} \tag{8}$$

all the coefficients $a_1, \ldots, a_k$ *being different from zero.*

Let us examine case (*b*) more carefully. There are two subcases: either (*b*1) where there are really no unknowns other than $x_1, \ldots, x_k$ or (*b*2) where there really are unknowns other than $x_1, \ldots, x_k$. In case (*b*1), our system of equations reduces to $a_1x_1 = b_1$, $a_2x_2 = b_2$, $\ldots$, $a_kx_k = b_k$, and, dividing, we find the unique solution $x_1 = b_1/a_1$, $\ldots$, $x_k = b_1/a_k$.

In case (*b*2), there are unknowns other than $x_1, \ldots, x_k$. Call the unknowns other than $x_1, \ldots, x_k$ by the letters $y_1, y_2, \ldots, y_n$. Since the coefficients $a_1, \ldots, a_k$ are different from zero, we can transpose and divide and write the system (8) of equations in the form

$$\begin{aligned} x_1 &= c_1 + d_1y_1 + e_1y_2 + \cdots + f_1y_n \\ x_2 &= c_2 + d_2y_1 + e_2y_2 + \cdots + f_2y_n \\ x_3 &= c_3 + d_3y_1 + e_3y_2 + \cdots + f_3y_n \\ &\cdots\cdots\cdots\cdots\cdots\cdots \\ x_k &= c_k + d_ky_1 + e_ky_2 + \cdots + f_ky_n. \end{aligned} \tag{9}$$

It is clear that this system of equations will be satisfied if we assign arbitrary values to the variables $y_1, \ldots, y_k$ and then determine the values of $x_1, \ldots, x_k$ from (9). In this case, our solution is evidently not unique.

Summarizing, we have the following theorem.

THEOREM

Let a system of arbitrarily many linear equations in arbitrarily many unknowns be given. If the process of elimination by subtraction described in the present section is repeatedly applied to the given system of linear equations, and carried through to the end, then we will arrive at (*a*), (*b*), or (*c*):

(*a*) A number of palpable contradictions of the form $0 = b$, b being some nonzero real number. In this case, the original system of equations has no solutions.

(*b*) An equivalent system of the form $a_1x_1 = b_1$, $a_2x_2 = b_2$, $\ldots$, $a_kx_k = b_k$, one for each of the unknowns in the original system of equations. In this case, there is a unique solution, and it is found by carrying the elimination process through to the end.

(*c*) An equivalent system of the form

$$\begin{aligned}
x_1 &= c_1 + d_1y_1 + e_1y_2 + \cdots + f_1y_n \\
x_2 &= c_2 + d_2y_1 + e_2y_2 + \cdots + f_2y_n \\
x_3 &= c_3 + d_3y_1 + e_3y_2 + \cdots + f_3y_n \\
&\cdots\cdots\cdots\cdots\cdots\cdots\cdots\cdots \\
x_k &= c_k + d_ky_1 + e_ky_2 + \cdots + f_ky_n,
\end{aligned}$$

the unknowns of the initial system being $x_1, \ldots, x_k$ and $y_1, \ldots, y_n$. In this case, the solution is not unique, but we obtain any solution by giving arbitrary values to the unknowns $y_1, \ldots, y_n$ and then determining the remaining unknowns from the equations (9).

Thus the question of solving systems of linear equations in arbitrarily many unknowns is settled in all possible cases.

EXERCISES

1. Find the solutions, if any, of the system of equations

$$\begin{aligned} 2v + x + y + z &= 0 \\ v - x + 2y + z &= 0 \\ 4v - x + 5y + 3z &= 1 \\ v - x + y - z &= 2. \end{aligned}$$

2. Find the solutions, if any, of the system of equations

$$\begin{aligned} x + y + z &= 1 \\ x - y - 2z &= 0 \\ x + 2y + 3z &= 1 \\ 3x - y - 5z &= 1. \end{aligned}$$

3. Find the solutions, if any, of the system of equations

$$\begin{aligned} v + 2x + y + z &= 0 \\ -v + x + 2y + z &= 0 \\ -v + 4x + 5y + 3z &= 1. \end{aligned}$$

4. Find the solutions, if any, of the system of equations

$$\begin{aligned} v + x + y + 2z &= 1 \\ v - x - y + 2z &= 2 \\ v + 2x - y + 2z &= 0 \\ v - 3x - 3y - 7z &= 4. \end{aligned}$$

(The teacher may assign additional exercises of this sort.)

Chapter 5

SPECIAL MATRICES OF PARTICULAR INTEREST

5-1. The Complex Numbers as Real Matrices

It is a general algebraic fact that many different kinds of algebraic systems can be expressed in terms of special collections of matrices. Many theorems of this kind are proved in modern higher algebra. Without attempting any such proof, we shall aim in the present chapter at showing how a number of interesting algebraic systems can be expressed in terms of matrices.

Suppose that we consider the set of all 2×2 matrices which have the form

$$\begin{bmatrix} a & b \\ -b & a \end{bmatrix}$$

i.e., have the form

$$aI + bJ,$$

where

$$J = \begin{bmatrix} 0 & 1 \\ -1 & 0 \end{bmatrix}.$$

The sum of two such matrices is of the same form:

$$(aI + bJ) + (a'I + b'J) = (a + a')I + (b + b')J. \qquad (1)$$

If we compute the square of J, we find that $J^2 = -I$. Hence it follows that the product of two such matrices has the same form:

$$(aI + bJ)(a'I + b'J) = (aa' - bb')I + (ab' + a'b)J. \quad (2)$$

It will probably have struck you by now that these matrices may be added and multiplied in just the same way as complex numbers. Indeed, the matrix J satisfies $J^2 = -I$; i.e., it is the square root of the negative of the unit matrix; and if we let the matrix

$$\begin{bmatrix} a & b \\ -b & a \end{bmatrix} = aI + bJ$$

correspond to the complex number $a + bi$, then equations (1) and (2) tell us that the matrices of our family are to be added and multiplied in exactly the same way as the corresponding complex numbers. In the terminology of abstract algebra, one says that our family of matrices and the family of complex numbers are *isomorphic*, that is, the two systems are merely different ways of representing the same algebra.

Thus we have represented the complex numbers as a family of real matrices. If we knew only about real numbers, and had never heard of complex numbers, we could have defined the complex numbers to be this set of real matrices, instead of defining them as ordered pairs of real numbers. The point is that, in searching for a square root of -1, there is no reason why, passing to 2×2 matrices, we could not use the matrix

$$J = \begin{bmatrix} 0 & 1 \\ -1 & 0 \end{bmatrix},$$

which we know satisfies $J^2 = -I$.

It is interesting to verify the commutative and division properties of complex numbers directly in terms of our family of matrices. Since

$$(aI + bJ)(a'I + b'J) = (aa' - bb')I + (ab' + a'b)J$$

and

$$(a'I + b'J)(aI + bJ) = (a'a - b'b)I + (b'a + ba')J,$$

the commutative property is obvious.

If we compute the product

$$\begin{bmatrix} a & -b \\ b & a \end{bmatrix}\begin{bmatrix} a & b \\ -b & a \end{bmatrix},$$

we find the result

$$\begin{bmatrix} a^2 + b^2 & 0 \\ 0 & a^2 + b^2 \end{bmatrix}.$$

Hence, if

$$\begin{bmatrix} a & b \\ -b & a \end{bmatrix}$$

is not zero, it has the reciprocal

$$\frac{1}{a^2 + b^2}\begin{bmatrix} a & -b \\ b & a \end{bmatrix}.$$

This verifies the division property of our family of matrices.

EXERCISE

Show that if we let the matrix

$$\begin{bmatrix} a & b \\ -b & a \end{bmatrix}$$

correspond to the complex number $a + bi$, then the matrix

$$\begin{bmatrix} a & -b \\ b & a \end{bmatrix}$$

corresponds to the complex conjugate $a - bi$.

5-2. The Quaternion Matrices

Consider the four matrices:

$$I = \begin{bmatrix} 1 & 0 & 0 & 0 \\ 0 & 1 & 0 & 0 \\ 0 & 0 & 1 & 0 \\ 0 & 0 & 0 & 1 \end{bmatrix}$$

$$J = \begin{bmatrix} 0 & 1 & 0 & 0 \\ -1 & 0 & 0 & 0 \\ 0 & 0 & 0 & -1 \\ 0 & 0 & 1 & 0 \end{bmatrix}$$

$$K = \begin{bmatrix} 0 & 0 & 1 & 0 \\ 0 & 0 & 0 & 1 \\ -1 & 0 & 0 & 0 \\ 0 & -1 & 0 & 0 \end{bmatrix}$$

$$L = \begin{bmatrix} 0 & 0 & 0 & 1 \\ 0 & 0 & -1 & 0 \\ 0 & 1 & 0 & 0 \\ -1 & 0 & 0 & 0 \end{bmatrix}.$$

If we compute their products, we find that they satisfy the multiplication rules

$$JK = -KJ = L; \qquad KL = -LK = J; \qquad LJ = -JL = K.$$
$$J^2 = K^2 = L^2 = -I.$$

Hence, if we consider the family of all matrices

$$aI + bJ + cK + dL = \begin{bmatrix} a & b & c & d \\ -b & a & -d & c \\ -c & d & a & b \\ -d & -c & -b & a \end{bmatrix},$$

we find that the sum of any two matrices of the given form has the same form:

$$(aI + bJ + cK + dL) + (a'I + b'J + c'K + d'L)$$
$$= (a + a')I + (b + b')J + (c + c')K + (d + d')L$$

and that the product of any two such matrices has also the same form:

$$(aI + bJ + cK + dL)(a'I + b'J + c'K + d'L)$$
$$= (aa' - bb' - cc' - dd')I + (ab' + ba' + cd' - dc')J$$
$$+ (ac' + ca' + db' - bd')K + (ad' + da' + bc' - cb')L.$$

The family of all matrices of the indicated form is called the family of *quaternion matrices* or simply the *quaternions*. If we put $a = a'$, $b = -b'$, $c = -c'$, and $d = -d'$, this formula specializes in an interesting way, since the coefficients of J, K, L vanish, and we find

$$(aI - bJ - cK - dL)(aI + bJ + cK + dL)$$
$$= (a^2 + b^2 + c^2 + d^2)I.$$

This proves that, if $aI + bJ + cK + dL$ is different from zero, then it has the reciprocal matrix

$$\frac{1}{a^2 + b^2 + c^2 + d^2}(aI - bJ - cK - dL).$$

Thus the quaternion matrices form a family of matrices having the following properties:

(*a*) The zero matrix is in the family.

(*b*) The sum of two matrices in the family is in the family.

(*c*) The product of two matrices in the family is in the family.

(*d*) Every nonzero matrix in the family has a reciprocal which is in the family.

But

(*e*) The commutative law is not satisfied.

Such an algebraic system is called a *skew field*.

Because the quaternions are so similar in their properties to the complex numbers (except that the commutative law breaks down), they are sometimes called a system of *hypercomplex numbers*.

5-3. Matrices with Complex Entries

Till now we have required all the entries in a matrix to be real numbers. There was no reason to do so; we could as well have allowed the entries in a matrix to be complex numbers. It is true that, in proving the various algebraic laws governing matrices, we used the properties of real numbers in the proof. But, as you can see by going back over all the proofs in the preceding sections, we never used any property of real numbers which was not equally well a property of complex numbers. Thus all our algebraic definitions can be made, and all our algebraic theorems can be proved, as well for matrices with complex entries as for matrices with real entries.

For some purposes, it is quite important to work in the algebra of all matrices with complex entries, rather than in the algebra of all matrices with real entries.

EXERCISES

1. What is

$$\begin{bmatrix} \frac{1}{2} & \frac{1}{3} & \frac{1-i}{4} \\ \frac{1+2i}{5} & \frac{1+3i}{6} & 1-i \\ 0 & 1+i & 2+i \end{bmatrix} + \begin{bmatrix} 1-i & 0 & i \\ -i & 1-i & 2i \\ 0 & i & 1-i \end{bmatrix}?$$

2. Solve the equation

$$2\left(4X + (1+i)X - \begin{bmatrix} 0 & i & 0 \\ -i & 0 & 1 \\ 1 & 0 & 1-i \end{bmatrix}\right) = \begin{bmatrix} 0 & 1+i & 1+2i \\ 1+3i & 4 & 0 \\ i & -i & 0 \end{bmatrix}$$

for the unknown matrix X.

3. Calculate the product

$$\begin{bmatrix} 0 & 1+i & 0 \\ -1 & 2-i & 0 \\ 0 & i & 0 \end{bmatrix} \begin{bmatrix} 2-i & 2+i & 4 \\ 0 & -3i & 0 \\ 5 & 1-i & 0 \end{bmatrix}.$$

4. The matrices

$$\sigma_1 = \begin{bmatrix} i & 0 \\ 0 & -i \end{bmatrix}, \qquad \sigma_2 = \begin{bmatrix} 0 & 1 \\ -1 & 0 \end{bmatrix}, \qquad \sigma_3 = \begin{bmatrix} 0 & i \\ i & 0 \end{bmatrix},$$

which are known as the *Pauli spin matrices*, play an extremely important role in atomic physics, where they are used to describe the spin of the electron around its axis. Prove that these matrices satisfy

$$\sigma_1^2 = {}_2^2 = \sigma_3^2 = -I$$

and

$$\sigma_1\sigma_2 = -\sigma_2\sigma_1 = \sigma_3; \qquad \sigma_2\sigma_3 = -\sigma_3\sigma_2 = \sigma_1; \qquad \sigma_3\sigma_1 = -\sigma_1\sigma_3 = \sigma_2.$$

These equations play a very important role in atomic physics.

5. Find a quadratic equation (with complex numbers as coefficients) satisfied by the matrix

$$A = \begin{bmatrix} 1+i & -i \\ i & 1-i \end{bmatrix}.$$

Use this to calculate the reciprocal A^{-1} and to calculate the power A^8.

6. Find the least equation (with complex number coefficients) satisfied by the matrix

$$A = \begin{bmatrix} 1+i & 0 & -i \\ 0 & 1-i & 1 \\ 1 & 1 & i \end{bmatrix}.$$

Does this matrix have a reciprocal? If so, what is its reciprocal?

7. Find the least equation (with complex number coefficients) satisfied by the matrix

$$A = \begin{bmatrix} 1+i & 0 & -i \\ 0 & 1 & 1+i \\ 2+2i & 1 & 1-i \end{bmatrix}.$$

Does this matrix have a reciprocal? If so, what is its reciprocal?

8. Solve the system of equations

$$\begin{aligned} (1+i)x + 2y - iz &= 1 \\ (1-i)x + y - (1-i)z &= 0 \\ (1-i)x + 2iy + (1+i)z &= 0 \end{aligned}$$

by the method of matrix inverses.

9. Solve the system of equations

$$\begin{aligned} u + (1+i)x + 2y + iz &= 1+i \\ iu + (1+i)x - 2iy + z &= 0 \\ (1-i)u + x - y - iz &= 0 \\ (1+i)u + 2ix - 2iy + (3+i)z &= 2-i \end{aligned}$$

by the method of elimination.

Chapter 6

MORE ALGEBRA OF MATRICES AND VECTORS

6-1. The Transpose Matrix

There exist a number of simple, useful, and theoretically important operations on matrices which we have not yet studied. A number of these will be studied in the present chapter.

The first of these operations is called taking the *transpose matrix*, and is described by the following definition.

DEFINITION

Let A be a matrix. Then the matrix B obtained from A by interchanging its rows and columns, i.e., the matrix B whose entries are defined by

$$[B]_{i,j} = [A]_{j,i},$$

is called the *transpose matrix* of A and written A^t.

Geometrically, B is obtained from A by "flipping over around the main diagonal," as in Fig. 1. The operation of taking the transpose matrix has a number of important properties, stated in the following theorem.

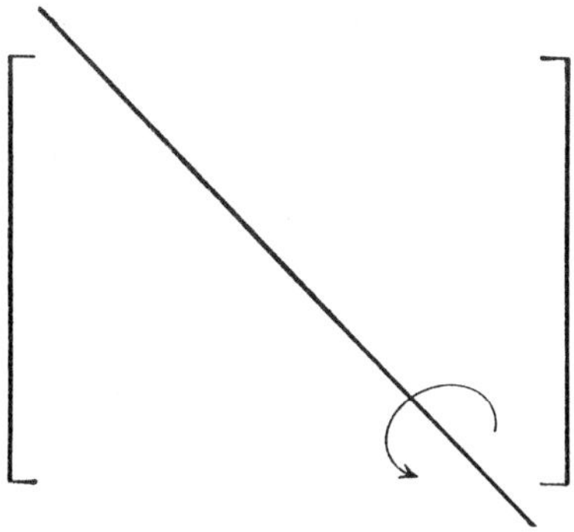

FIG. 1

THEOREM

Let A and B be two $n \times n$ matrices, and let x be a number. Then

(*a*) $(A + B)^t = A^t + B^t$ (the sum of the transposes is the transpose of the sum).

(*b*) $(xA)^t = xA^t$.

(*c*) $(AB)^t = B^tA^t$ (the transpose of the product is the product of the transposes taken in the reverse order).

(*d*) $(A^t)^t = A$ (the transpose of the transpose is the original matrix).

PROOF

The entry in the ith row and jth column of $(A + B)^t$ is the entry in the jth row and ith column of $A + B$. By definition of the sum of $A + B$, this is the sum of $[A]_{j,i} + [B]_{j,i}$, i.e., this is the sum of the entry in the ith row, jth column of the transpose A^t and the entry in the ith row, jth column of the transpose B^t. That is, taking all in all, the entry in the ith row, jth column of $(A + B)^t$ is the sum of the entry in the ith row, jth column of A^t and the entry in the ith row, jth column of B^t. Hence

$$(A + B)^t = A^t + B^t.$$

This proves (*a*).

If we use our notation more freely, we can write the proof in two lines:

$$[(A + B)^t]_{i,j} = [A + B]_{j,i} = [A]_{j,i} + [B]_{j,i} = [A^t]_{i,j} + [B^t]_{i,j},$$

so that $(A + B)^t = A^t + B^t$.

Using our notation in the same way, it is not hard to prove (*c*). We have

$$\begin{aligned}
[(AB)^t]_{i,j} &= [AB]_{j,i} \\
&= [A]_{j,1}[B]_{1,i} + [A]_{j,2}[B]_{2,i} + \cdots + [A]_{j,n}[B]_{n,i} \\
&= [A^t]_{1,j}[B^t]_{i,1} + [A^t]_{2,j}[B^t]_{i,2} + \cdots + [A^t]_{n,j}[B^t]_{i,n} \\
&= [B^t]_{i,1}[A^t]_{1,j} + [B^t]_{i,2}[A^t]_{2,j} + \cdots + [B^t]_{i,j}[A^t]_{n,i} \\
&= [B^tA^t]_{i,j}.
\end{aligned}$$

Hence $(AB)^t = B^tA^t$, proving (*c*).

We shall not give the proofs of (*b*) or (*d*) but will leave them to the student as an exercise.

EXERCISES

1. Prove part (*b*) of the theorem of the preceding section.

2. Prove part (*d*) of the theorem of the preceding section.

3. Write down the transposes of the matrices

$$\begin{bmatrix} 1 & 2 & 3 \\ 4 & 5 & 6 \\ 7 & 8 & 9 \end{bmatrix} \quad \text{and} \quad \begin{bmatrix} 1 & 2 & 3 & 4 \\ 5 & 6 & 7 & 8 \\ 9 & 10 & 11 & 12 \\ 13 & 14 & 15 & 16 \end{bmatrix}.$$

4. Suppose that a matrix A satisfies $A^t = -A$. Which of its entries must be zero? Give an example of such a 3×3 matrix.

6-2. The Trace of a Matrix

DEFINITION

Let A be an $n \times n$ matrix. Then the trace of the matrix A, written tr (A), is the numerical sum

$$\text{tr}\,(A) = [A]_{1,1} + [A]_{2,2} + \cdots + [A]_{n,n}.$$

That is, the trace is the sum of the entries in first row and first column, second row and second column, etc. Thus, for instance, the trace

$$\operatorname{tr}\left(\begin{bmatrix} 1 & 0 & 0 & 0 \\ 0 & 1 & 0 & 0 \\ 0 & 0 & 1 & 0 \\ 0 & 0 & 0 & 1 \end{bmatrix}\right)$$

of the 4×4 unit matrix is $1 + 1 + 1 + 1 = 4$. The trace has a number of important properties, which will be described in the present section.

In the Σ notation the trace of an $n \times n$ matrix A may be written

$$\operatorname{tr}(A) = \sum_{i=1}^{n} [A]_{i,i}.$$

It should be noted that the trace is a number, so that $\operatorname{tr}(A)$ is a numerical function of the matrix A. The following theorem gives the main properties of the trace.

THEOREM

Let A and B be two $n \times n$ matrices, and let x be a number. Then

(*a*) $\operatorname{tr}(A^t) = \operatorname{tr}(A)$
(*b*) $\operatorname{tr}(A + B) = \operatorname{tr}(A) + \operatorname{tr}(B)$
(*c*) $\operatorname{tr}(xA) = x \operatorname{tr}(A)$
(*d*) $\operatorname{tr}(AB) = \operatorname{tr}(BA)$.

PROOF

Since the entry $[A^t]_{i,i}$ in the ith row, ith column of A^t is the same as the entry $[A]_{i,i}$, (*a*) is clear.

Parts (*b*) and (*c*) are easy and will be left to the student as an exercise.

To prove (*d*), we argue as follows.

The entry $[AB]_{i,i}$ is, by definition,

$$[A]_{i,1}[B]_{1,i} + [A]_{i,2}[B]_{2,i} + \cdots + [A]_{i,n}[B]_{n,i};$$

i.e., it is the sum of all the terms $[A]_{i,j}[B]_{j,i}$ with a fixed value of i. If we sum over i to get tr (AB), we get the sum of all terms $[A]_{i,j}[B]_{j,i}$, i and j ranging freely between 1 and n.

The entry $[BA]_{j,j}$ is, by definition,

$$[B]_{j,1}[A]_{1,j} + [B]_{j,2}[A]_{2,j} + \cdots + [B]_{j,n}[A]_{n,j};$$

i.e., it is the sum of all the terms $[B]_{j,i}[A]_{i,j}$ with a fixed value of j. If we sum over j to get tr (BA), we get the sum of all terms $[B]_{j,i}[A]_{i,j}$, i and j ranging freely between 1 and n. Hence it is clear that tr (AB) = tr (BA). This proves (d). Q.E.D.

There is an interesting connection between the circle product of two vectors and the trace. This is described in the following theorem.

THEOREM

Let U and V be vectors of size n. Then

$$\text{tr } (U^tV) = U \circ V.$$

PROOF

The product U^tV of the vector V by the matrix U is a vector. Hence all the entries in its second, third, etc., columns vanish. The transpose matrix $(U^tV)^t = V^t(U^t)^t = V^tU$ is the product of the vector U by the matrix V^t and, consequently, is also a vector. Thus all the entries in the second, third, etc., columns of the transpose of U^tV vanish. Since we get the transpose matrix by interchanging rows and columns, it is clear that all the entries in the second, third, etc., *rows* of U^tV vanish. Thus, taking all in all, the only possible nonzero entry in U^tV is $[U^tV]_{1,1}$. This makes it clear that tr $(U^tV) = [U^tV]_{1,1}$. But

$$\begin{aligned}[U^tV]_{1,1} &= [U^t]_{1,1}[V]_{1,1} + [U^t]_{1,2}[V]_{2,1} + \cdots + [U^t]_{1,n}[V]_{n,1} \\ &= [U]_{1,1}[V]_{1,1} + [U]_{2,1}[V]_{2,1} + \cdots + [U]_{n,1}[V]_{n,1}.\end{aligned}$$

Thus tr (U^tV) is the sum of the product of each element in the first column of U by the corresponding element in the first column of V. If we write U and V in the notation $U = [u_1, \ldots, u_n]$ and $V = [v_1, \ldots, v_n]$, this may be written as

$$\text{tr}\ (U^tV) = u_1v_1 + u_2v_2 + \cdots + u_nv_n.$$

But this is exactly the expression which defined the circle product $U \circ V$. Hence tr $(U^tV) = U \circ V$. Q.E.D.

Much of the importance of the transpose matrix comes from its connection with the circle product of two vectors. This connection is described in the following theorem.

THEOREM

Let U and V be vectors, and let A be a matrix, all the same size. Then

$$(AU) \circ V = U \circ (A^tV).$$

PROOF

By the two preceding theorems, and by the next-to-last theorem of Section 4-7,

$$\begin{aligned} AU \circ V = V \circ AU = \text{tr}\ (V^t(AU)) = \text{tr}\ (AUV^t) &= \text{tr}\ ((AUV^t)^t) \\ &= \text{tr}\ ((V^t)^tU^tA^t) \\ &= \text{tr}\ (A^tVU^t) \\ &= A^tV \circ U \\ &= U \circ A^tV. \end{aligned}$$

Q.E.D.

EXERCISES

1. Prove part (*b*) of the first theorem of the preceding section.
2. Prove part (*c*) of the first theorem of the preceding section.
3. Calculate the trace of the matrix

$$\begin{bmatrix} 1 & 0 & 0 & 1 \\ 1 & 1 & 0 & 0 \\ 1 & 1 & 1 & 0 \\ 0 & -1 & -1 & -1 \end{bmatrix}.$$

4. Let U and V be two vectors of the same size. Prove that

$$U \circ V = \text{tr } (U^tV) = \text{tr } (VU^t) = \text{tr } (V^tU) = \text{tr } (UV^t).$$

5. Prove that $\text{tr } (AA^t) \geq 0$ and that $\text{tr } (AA^t) > 0$ unless $A = 0$.

6. Prove the inequality

$$|\text{tr } (AB)| \leq \{\text{tr } (AA^t)\}^{\frac{1}{2}}\{\text{tr } (BB^t)\}^{\frac{1}{2}}.$$

(*Hint:* Follow the proof of the corresponding inequality for the circle product of two vectors.)

6-3. The Cross Product of Matrices

The failure of the commutative law permits us to define the following interesting product between two matrices.

DEFINITION

Let A and B be two matrices of the same size. Then the expression $AB - BA$ is called the cross product of A and B and is written $A \times B$.

A number of interesting algebraic properties of the cross product are stated in the following theorem.

THEOREM

Let A, B, C be matrices all of the same size, and let x be a number. Then

(*a*) $A \times B = -(B \times A)$
(*b*) $(xA) \times B = A \times (xB) = x(A \times B)$
(*c*) $A \times (B + C) = A \times B + A \times C$
(*d*) $(B + C) \times A = B \times A + C \times A$
(*e*) $A \times (B \times C) = B \times (A \times C) + C \times (B \times A)$.

Statement (*a*) shows that for the cross product the commutative law breaks down very violently: it is replaced by an *anticommuta-*

tive law. Statement (e) shows that even the associative law $x(yz) = (xy)z$ breaks down for the cross product: in dealing with cross products, we must be careful how we place our parentheses.

PROOF

All the statements of the theorem follow readily from the definition of the cross product.

We have

$$A \times B = AB - BA = -(BA - AB) = -(B \times A).$$

This proves (a).

We have

$$\begin{aligned} A \times (B \times C) &= A(B \times C) - (B \times C)A \\ &= A(BC - CB) - (BC - CB)A \\ &= ABC - ACB - BCA + CBA \end{aligned}$$

while

$$\begin{aligned} B \times (A \times C) &+ C \times (B \times A) \\ &= B(A \times C) - (A \times C)B \\ &\qquad + C(B \times A) - (B \times A)C \\ &= B(AC - CA) - (AC - CA)B \\ &\qquad + C(BA - AB) - (BA - AB)C \\ &= BAC - BCA - ACB + CAB + CBA - CAB \\ &\qquad - BAC + ABC \\ &= CBA - BCA - ACB + ABC. \end{aligned}$$

Thus $A \times (B \times C) = B \times (A \times C) + C \times (B \times A)$, which proves (e).

Parts (b), (c), and (d) will not be proved, but their proofs will be left to the student as an exercise. Q.E.D.

An algebraic system with the properties (a) to (e) is often called a *Lie algebra.*

A certain special sort of matrix has a simple but interesting relation to the cross product. The following definitions and theorem concern this point.

DEFINITION

A matrix A such that $A^t = -A$ is called a *skew matrix*.

THEOREM

Let A and B be skew matrices of the same size. Then $A \times B$ is a skew matrix.

PROOF

Since $A^t = -A$, $B^t = -B$, we have

$$\begin{aligned}(A \times B)^t = (AB - BA)^t \\ &= (B^tA^t - A^tB^t) = (BA - AB) = B \times A \\ &= -A \times B.\end{aligned}$$

Q.E.D.

EXERCISES

1. Prove part (*b*) of the first theorem of this section.
2. Prove part (*c*) of the first theorem of this section.
3. Prove part (*d*) of the first theorem of this section.
4. Prove that the trace of a skew matrix is zero.
5. Prove that if $A^t = A$, $B^t = B$ then $A \times B$ is a skew matrix.

6-4. 3 × 3 Skew Matrices and Vectors of Size 3

There is an interesting connection between skew matrices of size 3 and vectors of size 3, which it is the aim of the present section to exploit. A skew matrix A satisfies the equation $[A]_{i,j} = -[A]_{j,i}$. Thus $[A]_{i,i} = -[A]_{i,i}$, so that $[A]_{i,i} = 0$. A 3×3 skew matrix must consequently have the following form:

$$A = \begin{bmatrix} 0 & u & v \\ -u & 0 & w \\ -v & -w & 0 \end{bmatrix}.$$

Clearly, whatever the value of u, v, w, the matrix displayed above is a skew matrix. It is consequently natural to make the following definition.

DEFINITION

The skew matrix

$$A = \begin{bmatrix} 0 & u & v \\ -u & 0 & w \\ -v & -w & 0 \end{bmatrix}$$

is said to be the *skew matrix of the vector* $[u,v,w]$; and the vector $[u,v,w]$ is said to be the vector of the skew matrix A.

This definition establishes a correspondence between skew matrices of size 3 and vectors of size 3. It is clear that each skew matrix is the skew matrix of a unique vector and that each vector is the vector of a unique skew matrix.

It is interesting to see how various properties of vectors can be represented in terms of the skew matrices of these vectors. This is done in the next theorem.

THEOREM

Let U and V be vectors of size 3, and let A and B be the skew matrices of U and V, respectively. Let x be a number. Then

(*a*) $A + B$ is the skew matrix of $U + V$

(*b*) xA is the skew matrix of xU

(*c*) $U \circ V = -\frac{1}{2} \operatorname{tr}(AB)$.

PROOF

Let $U = [u_1,u_2,u_3]$, $V = [v_1,v_2,v_3]$. Then

$$A = \begin{bmatrix} 0 & u_1 & u_2 \\ -u_1 & 0 & u_3 \\ -u_2 & -u_3 & 0 \end{bmatrix} \qquad B = \begin{bmatrix} 0 & v_1 & v_2 \\ -v_1 & 0 & v_3 \\ -v_2 & -v_3 & 0 \end{bmatrix}.$$

Then $U + V = [u_1 + v_1,\ u_2 + v_2,\ u_3 + v_3]$, and the skew matrix of this vector is

$$\begin{bmatrix} 0 & u_1 + v_1 & u_2 + v_2 \\ -(u_1 + v_1) & 0 & u_3 + v_3 \\ -(u_2 + v_2) & -(u_3 + v_3) & 0 \end{bmatrix},$$

clearly equal to $A + B$. This proves (*e*).

Parts (*b*) and (*c*) may be proved by equally elementary computations. The proofs will not be given, but will be left to the student as an exercise. Q.E.D.

We know from the final theorems of the preceding section that the cross product of two skew matrices is a skew matrix. The correspondence between vectors of size 3 and skew matrices of size 3 that we have established gives us an obvious method for defining a new product of two vectors. This is done in the following definition.

DEFINITION

Let U and V be two vectors of size 3, and let A and B be the skew matrices of U and V, respectively. Then the *cap product* of U and V is the vector of the skew matrix $A \times B$ and is written $U \wedge V$.

Thus, to take the cap product of two vectors U and V, we *first* take the skew matrices A and B of the vectors U and V, *then* form the cross product $A \times B$ of A and B, and finally take the vector W of the skew matrix $A \times B$. By definition, W is the cap product of U and V; $W = U \wedge V$.

It should be carefully noted that the cap product of two vectors of size 3 is itself a vector of size 3. In this regard, the cap product differs fundamentally from the circle product: the cap product of two vectors of size 3 is a vector; the circle product is a number. For this reason the cap product is sometimes called the *vector product* of two vectors.

It should also be noted that the cap product has been defined *only* for vectors of size 3. It will *not* be defined for vectors of any other size. The existence of a cap product is a special feature of vectors of size 3. Of course, since we live in a three-dimensional space, vectors of size 3 are exceptionally important. The radio engineer, the mathematical physicist, the aeronautical engineer all make great use of the cap product of vectors of size 3.

The final theorem of the preceding section gives the following properties of the cap product immediately.

THEOREM

Let U, V, and W be vectors all of size 3, and let x be a number. Then

(*a*) $U \wedge V = -V \wedge U$
(*b*) $(xU) \wedge V = U \wedge (xV) = x(U \wedge V)$
(*c*) $U \wedge (V + W) = U \wedge V + U \wedge W$
(*d*) $(V + W) \wedge U = V \wedge U + W \wedge U$
(*e*) $U \wedge (V \wedge W) = V \wedge (U \wedge W) + W \wedge (V \wedge U)$.

PROOF

All these statements follow at once from the final theorem of the preceding section, simply by use of the correspondence between vectors of size 3 and skew matrices of size 3 which has been established in the second definition above. Thus, to prove (*e*), we reason as follows: Let A, B, C be the skew matrices of U, V, W. Then, by definition, $B \times C$ is the skew matrix of $V \wedge W$, so that $A \times (B \times C)$ is the skew matrix of $U \wedge (V \wedge W)$. Similarly, $B \times (A \times C)$ is the skew matrix of $V \wedge (U \wedge W)$, and $C \times (B \times A)$ is the skew matrix of $W \wedge (V \wedge U)$. Since

$$A \times (B \times C) = B \times (A \times C) + C \times (B \times A)$$

by (*e*) of the final theorem of the preceding section, part (*e*) of the present theorem follows immediately from part (*e*) of the preceding theorem.

Parts (*a*), (*b*), (*c*), and (*d*) of the present theorem follow in the same way. The proofs will not be given, but will be left to the student as an exercise. Q.E.D.

Let $U = [u_1,u_2,u_3]$ and $V = [v_1,v_2,v_3]$ be two vectors of size 3.

Then the skew matrices of these vectors are

$$A = \begin{bmatrix} 0 & u_1 & u_2 \\ -u_1 & 0 & u_3 \\ -u_2 & -u_3 & 0 \end{bmatrix} \quad \text{and} \quad B = \begin{bmatrix} 0 & v_1 & v_2 \\ -v_1 & 0 & v_3 \\ -v_2 & -v_3 & 0 \end{bmatrix}.$$

If we work out the cross product $A \times B = AB - BA$ by explicitly computing AB and BA and then subtracting, we find

$$A \times B = \begin{bmatrix} 0 & u_2v_3 - v_2u_3 & u_3v_1 - v_3u_1 \\ v_2u_3 - u_2v_3 & 0 & u_1v_2 - v_1u_2 \\ v_3u_1 - u_3v_1 & v_1u_2 - u_1v_2 & 0 \end{bmatrix}.$$

(Check the computation yourself.)

This proves the following theorem.

THEOREM

The cap product of the vectors $U = [u_1,u_2,u_3]$ and $V = [v_1,v_2,v_3]$ is

$$U \wedge V = [u_2v_3 - v_2u_3,\ u_3v_1 - v_3u_1,\ u_1v_2 - v_1u_2].$$

This theorem gives us a convenient explicit formula for the cap product. We use it to prove the following two very useful theorems.

THEOREM

Let U and V be two vectors of size 3. The $U \wedge V = 0$ if and only if either

(*a*) $U = 0$ or

(*b*) V is a numerical multiple of U.

THEOREM

Let U, V, and W be three vectors of size 3. Then

$$U \wedge (V \wedge W) = (U \circ W)V - (U \circ V)W.$$

PROOF OF FIRST THEOREM

If $U = 0$, then $U = 0 \cdot U$, and

$$U \wedge V = (0 \cdot U) \wedge V = 0 \cdot (U \wedge V) = 0.$$

If $V = xU$, then $U \wedge V$ is equal to its own negative, and thus is zero. This shows that, if (a) or (b) holds, $U \wedge V = 0$.

Now suppose that $U \wedge V = 0$. We shall show that if $U \neq 0$ then V is a numerical multiple of U, and this will prove our theorem. Let $U = [u_1,u_2,u_3]$ and $V = [v_1,v_2,v_3]$. Since $U \neq 0$, one of its entries is different from zero. Suppose simply for the sake of definiteness that $u_1 \neq 0$; if we supposed instead that $u_2 \neq 0$ or $u_3 \neq 0$, the following argument could be given in just the same way. Put $v_1/u_1 = x$. Since by our explicit formula for the cap product $u_1v_2 - v_1u_2 = 0$, $u_1v_3 - v_1u_3 = 0$, it follows that $v_1 = xu_1$, $v_2 = xu_2$, $v_3 = xu_3$. Thus $V = xU$, so that V is a numerical multiple of U.

PROOF OF SECOND THEOREM

Let $U = [u_1,u_2,u_3]$, $V = [v_1,v_2,v_3]$, and $W = [w_1,w_2,w_3]$. Then $V \wedge W = [v_2w_3 - w_2v_3,\ v_3w_1 - w_3v_1,\ v_1w_2 - w_1v_2]$.

Applying our explicit formula for the cap product again, we find

$$\begin{aligned}
U \wedge (V \wedge W) &= [u_2(v_1w_2 - w_1v_2) - (v_3w_1 - w_3v_1)u_3, \\
&\quad u_3(v_2w_3 - w_2v_3) - (v_1w_2 - w_1v_2)u_1, \\
&\qquad u_1(v_3w_1 - w_3v_1) - (v_2w_3 - w_2v_3)u_2] \\
&= [v_1(u_2w_2 + u_3w_3) - w_1(u_2v_2 + u_3v_3), \\
&\quad v_2(u_1w_1 + u_3w_3) - w_2(u_1v_1 + u_3v_3), \\
&\qquad v_3(u_1w_1 + u_2w_2) - w_3(u_1v_1 + u_3v_3)] \\
&= [v_1(u_1w_1 + u_2w_2 + u_3w_3) \\
&\qquad - w_1(u_1v_1 + u_2v_2 + u_3w_3), \\
&\quad v_2(u_1w_1 + u_2w_2 + u_3w_3) \\
&\qquad - w_2(u_1v_1 + u_2v_2 + u_3v_3), \\
&\quad v_3(u_1w_1 + u_2w_2 + u_3w_3) \\
&\qquad - w_3(u_1v_1 + u_2v_2 + u_3w_3)] \\
&= (U \circ W)[v_1,v_2,v_3] - (U \circ V)[w_1,w_2,w_3] \\
&= (U \circ W)V - (U \circ V)W.
\end{aligned}$$

Q.E.D.

A number of other useful formulae for the cap product of two vectors can be proved by using various facts about the cross product of matrices. Some of these are given in the following theorem.

THEOREM

Let U, V, and W be three vectors of size 3. Then

(a) $U \circ (V \wedge W) = (U \wedge V) \circ W$
(b) $U \circ (U \wedge V) = 0$
(c) $(U \wedge V) \circ (U \wedge V) = (V \circ V)(U \circ U) - (U \circ V)^2$.

PROOF

Let A, B, and C be the skew matrices of U, V, and W, respectively. Then, by the definition of the cap product and by part (c) of the first theorem of this section, statement (a) is equivalent to the statement

$$\text{tr } (A(B \times C)) = \text{tr } ((A \times B)C).$$

Now $A(B \times C) = A(BC - CB) = ABC - ACB$, and

$$(A \times B)C = (AB - BA)C = ABC - BAC.$$

Thus statement (a) is equivalent to the statement

$$\text{tr } (ABC) - \text{tr } (ACB) = \text{tr } (ABC) - \text{tr } (BAC).$$

Since by the theorem of Section 6-2,

$$\text{tr } (ACB) = ((AC)B) = \text{tr } (B(AC)) = \text{tr } (BAC),$$

this is obvious. Hence (a) is proved.

Using (a), we have $U \circ (U \wedge V) = (U \wedge U) \circ V$. Since $U \wedge U = 0$ by the second theorem above, $U \circ (U \wedge V) = 0$.

Using (a), we have

$$(U \wedge V) \circ (U \wedge V)$$
$$= ((U \wedge V) \wedge U) \circ V = -(U \wedge (U \wedge V)) \circ V.$$

Then, using the preceding theorem, it follows that

$$\begin{aligned}(U \wedge V) \circ (U \wedge V) &= -((U \circ V)U - (U \circ U)V) \circ V \\ &= (V \circ V)(U \circ U) - (U \circ V)^2.\end{aligned}$$

This proves (*c*). Q.E.D.

EXERCISES

1. Prove part (*b*) of the first theorem of the preceding section.

2. Prove part (*c*) of the first theorem of the preceding section.

3. Prove part (*a*) of the second theorem of the preceding section.

4. Prove part (*b*) of the second theorem of the preceding section.

5. Prove part (*c*) of the second theorem of the preceding section.

6. Prove part (*d*) of the second theorem of the preceding section.

(A number of additional exercises on the algebraic properties of the circle and cap products are given at the end of the next section.)

6-5. Geometry of the Cap Product

We know from Chapter 4 that vectors of size 3 may be represented by directed segments $\overrightarrow{PQ}$. If the coordinates of P and Q are P_1, P_2, P_3 and Q_1, Q_2, Q_3, respectively, then the directed segment $\overrightarrow{PQ}$ represents the vector

$$V = [Q_1 - P_1, Q_2 - P_2, Q_3 - P_3].$$

The length of the segment $\overrightarrow{PQ}$ has been shown to be given by the formula

$$\|V\| = (v_1^2 + v_2^2 + v_3^2)^{\frac{1}{2}} = (V \circ V)^{\frac{1}{2}},$$

where v_1, v_2, and v_3 are the three entries of V. If $\overrightarrow{PR}$ is another directed segment making an angle θ with the segment $\overrightarrow{PQ}$ and $\overrightarrow{PR}$ represents the vector U, then we showed that

$$U \circ V = (\widehat{PR})(\widehat{PQ}) \cos \theta,$$

$\widehat{PR}$ and $\widehat{PQ}$ being the lengths of the directed segments $\overrightarrow{PR}$ and $\overrightarrow{PQ}$. Moreover, $U \circ V = 0$ if and only if $\overrightarrow{PQ}$ and $\overrightarrow{PR}$ are perpendicular. This formula gives the geometric significance of the circle product. The cap product of two vectors, which we defined and studied in the preceding section, has also an interesting geometric interpretation, and in the present section we intend to study this interpretation.

Let us first see when $U \wedge V = 0$. According to the theorems of the preceding section, this happens only either when $U = 0$ (so that $P = Q$) or when V is a numerical multiple of U. According to the theorems of Chapter 4, V is a numerical multiple of U if and only if the segment $\overrightarrow{PR}$ is collinear with the segment $\overrightarrow{PQ}$.

Thus, summarizing: $U \wedge V = 0$ if and only if either one of the directed segments $\overrightarrow{PQ}$ and $\overrightarrow{PR}$ reduces to a single point, or if these segments are collinear.

Let us now exclude this case and suppose from now on that $\overrightarrow{PQ}$ and $\overrightarrow{PR}$ are not collinear and that neither reduces to a single point. Then $U \wedge V$ is a nonzero vector and may, consequently, be represented by a directed segment $\overrightarrow{PS}$. What is the geometric description of this segment $\overrightarrow{PS}$?

We know from (*b*) of the last theorem of the preceding section that $U \circ (U \wedge V) = 0$ and that

$$V \circ (U \wedge V) = -V \circ (V \wedge U) = 0$$

also. Thus $\overrightarrow{PS}$ must be perpendicular both to $\overrightarrow{PQ}$ and $\overrightarrow{PR}$. Hence $\overrightarrow{PS}$ must lie along the line through P perpendicular to the plane of $\overrightarrow{PQ}$ and $\overrightarrow{PR}$. This determines the direction (up to reversal) of $\overrightarrow{PS}$.

How long is $\overrightarrow{PS}$? By part (*c*) of the final theorem of the preceding section,

$$\begin{aligned} \|V \wedge U\| &= (U \circ U)(V \circ U) - (U \circ V)^2)^{\frac{1}{2}} \\ &= (\widehat{PQ}^2\widehat{PR}^2 - (\widehat{PQ}\widehat{PR} \cos \theta)^2)^{\frac{1}{2}} \\ &= ((\widehat{PQ}^2\widehat{PR}^2)(1 - \cos^2 \theta))^{\frac{1}{2}} \\ &= \widehat{PQ}\widehat{PR}(\sin^2 \theta)^{\frac{1}{2}} \\ &= \widehat{PQ}\widehat{PR}|\sin \theta|. \end{aligned}$$

Thus, the length of $\overrightarrow{PS}$ is the product

$$\widehat{PQ}\widehat{PR}|\sin \theta|$$

of the lengths of $\overrightarrow{PQ}$, $\overrightarrow{PR}$ and the absolute value of the sine of the angle between $\overrightarrow{PQ}$ and $\overrightarrow{PR}$.

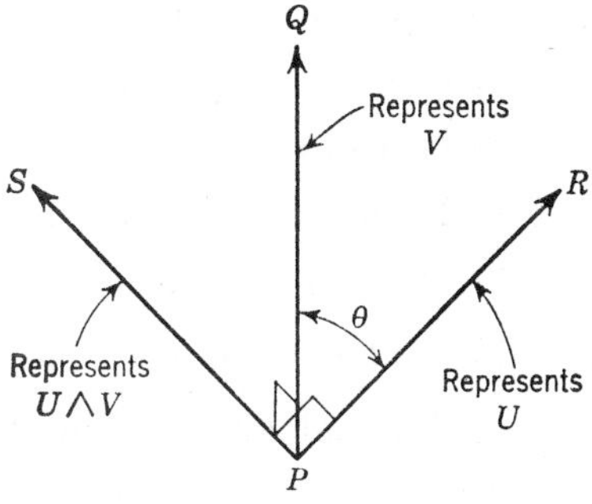

FIG. 2. The cap product.

Thus we have determined the line along which $\overrightarrow{PS}$ lies, and its length besides; but $\overrightarrow{PS}$ is not yet fully determined, since there are still two oppositely directed line segments which meet our specifications.

Which of these two oppositely directed line segments is the one representing $U \wedge V$?

We may ask this question of *orientation* as follows. Suppose that we take our right hand and place the middle finger along the directed segment $\overrightarrow{PR}$ representing U and the thumb along the directed segment $\overrightarrow{PQ}$ representing V. Then does the segment $\overrightarrow{PS}$ point into the same half-space as the index finger or in the opposite direction? When we answer this question, we will have determined the segment $\overrightarrow{PS}$ completely.

Now, we may answer our question as follows. Keeping $\overrightarrow{PQ}$ fixed, gradually rotate $\overrightarrow{PR}$ in the plane of PRQ until it comes into perpendicularity with $\overrightarrow{PQ}$ without passing through the line of $\overrightarrow{PQ}$. During this process, the segment $\overrightarrow{PS}$ will gradually change in length, always lying along the same line. Since $\overrightarrow{PR}$ and $\overrightarrow{PQ}$ never become collinear, the length of $\overrightarrow{PS}$ never becomes zero; that is, the point S never passes through the plane of PQR. Hence the half-space into which $\overrightarrow{PS}$ points does not change.

Now that the three segments $\overrightarrow{PQ}$, $\overrightarrow{PR}$, and $\overrightarrow{PS}$ have been brought into a position in which all three are perpendicular, we gradually move $\overrightarrow{PQ}$ and $\overrightarrow{PR}$, keeping them perpendicular, till P lies at the origin of coordinates, $\overrightarrow{PR}$ pointing along the x axis in the positive direction and $\overrightarrow{PQ}$ pointing along the axis in the negative direction. As $\overrightarrow{PQ}$ and $\overrightarrow{PR}$ are moved, the entries in the vectors V and U which they represent change gradually, and hence the entries in $U \wedge V$ change gradually. Thus the directed segment $\overrightarrow{PS}$ representing $U \wedge V$ changes gradually. Since $\overrightarrow{PR}$ and $\overrightarrow{PQ}$ are never parallel, $U \wedge V$ is never zero, so that the point S never passes through the plane of $\overrightarrow{PQ}$ and $\overrightarrow{PR}$. Thus, the directed segment $\overrightarrow{PS}$ always points into the same half-space of the plane of $\overrightarrow{PQ}$ and $\overrightarrow{PR}$. To explain this somewhat differently, if $\overrightarrow{PR}$, $\overrightarrow{PQ}$, and $\overrightarrow{PS}$ are initially oriented as middle finger, thumb, and index finger of the right hand, then they preserve this relative orientation. If $\overrightarrow{PS}$ is initially oriented in the reverse direction (so that $\overrightarrow{PR}$, $\overrightarrow{PQ}$, and $\overrightarrow{PS}$ are initially oriented as middle finger, thumb, and index finger of the *left* hand), then they preserve this reverse relative orientation.

But in the final position, $\overrightarrow{PR}$ points in the positive x direction and hence represents a vector $[x,0,0]$ with x positive; $\overrightarrow{PQ}$ points in the positive y direction and hence represents a vector $[0,y,0]$ with y positive. Using the formula for the cap product, we find that $U \wedge V$ is the vector $[0,0,xy]$, so that $\overrightarrow{PS}$ must point in the positive direction. Thus $\overrightarrow{PR}$, $\overrightarrow{PQ}$, and $\overrightarrow{PS}$ have the orientations of the positive x, y, and z axes, i.e., of the middle finger, thumb, and index finger of the *right* hand. Since they have this relative orientation in their final position, it follows by what has been said above that they have the same orientation in their initial position.

Summarizing, we see that we have proved the following theorem.

THEOREM

Let $\overrightarrow{PR}$ and $\overrightarrow{PQ}$ be two directed segments representing vectors U and V, respectively. Then

(a) The cap product $U \wedge V$ is zero if and only if either one of the two directed segments reduces to a single point, or both are collinear.

(b) If $U \wedge V \neq 0$, then the vector $U \wedge V$ is represented by the directed segment PS determined as follows:

(1) $\overrightarrow{PS}$ is of the length

$$\widehat{PS} = \widehat{PR}\widehat{PQ}|\sin \theta|,$$

$\widehat{PR}$ and $\widehat{PQ}$ being the lengths of the segments $\overrightarrow{PR}$ and $\overrightarrow{PQ}$ and θ the angle between them.

(2) $\overrightarrow{PS}$ is perpendicular to the common plane of $\overrightarrow{PR}$ and $\overrightarrow{PQ}$.

(3) The three directed segments $\overrightarrow{PR}$, $\overrightarrow{PQ}$, $\overrightarrow{PS}$ have the relative

orientation of the middle finger, thumb, and index finger of the right hand.

Part (3) of this theorem is sometimes called the right-hand rule for determining the direction of the segment representing a cap product.

EXERCISES

1. Let the directed segments $\overrightarrow{PQ}$ and $\overrightarrow{RS}$ represent the vectors U and V. Then $U \wedge V = 0$ if and only if the lines PQ and RS are either identical or parallel.

2. If $\overrightarrow{PQ}$, $\overrightarrow{PR}$, and $\overrightarrow{PS}$ are three oriented segments representing vectors U, V, and W, respectively, and if $(U \wedge V) \circ W = 0$, then $\overrightarrow{PQ}$, $\overrightarrow{PR}$, and $\overrightarrow{PS}$ are coplanar, and vice versa.

3. Prove the vector identities

$$(U \wedge V) \circ (U \wedge W) = (U \circ U)(V \circ W) - (U \circ V)(U \circ W).$$
$$(U \wedge V) \circ (W \wedge X) = (U \circ W)(V \circ X) - (U \circ X)(V \circ W).$$

4. If three oriented segments $\overrightarrow{PQ}$, $\overrightarrow{PR}$, and $\overrightarrow{PS}$ are the edges of a parallelepiped Π, and $\overrightarrow{PQ}$, $\overrightarrow{PR}$, and $\overrightarrow{PS}$ represent the vectors U, V, and W, respectively, then the volume of Π is the absolute value of the expression $(U \wedge V) \circ W$.

5. Prove the vector identity

$$[(V \wedge W) \wedge (W \wedge U)] \circ (U \wedge V) = ((U \wedge V) \circ W)^2.$$

Chapter 7

EIGENVALUES AND EIGENVECTORS

7-1. Matrix Reciprocals and Vectors

Let A be an $n \times n$ matrix and V be a vector. Let $AV = 0$. Then if A has a reciprocal A^{-1}, it follows that $V = A^{-1}0 = 0$. Thus, if A has a reciprocal A^{-1}, it follows that the equation $AV = 0$ has no vector solution other than the vector zero.

It is an important fact that the converse of this statement is true. This fact is stated in the following theorem.

THEOREM

Let A be an $n \times n$ matrix. Then a necessary and sufficient condition that A have a reciprocal is that there exist no nonzero vector such that $AV = 0$.

PROOF

We have already proved the "necessity" and have only to prove that, if A has no reciprocal, there exists a nonzero vector V such that $AV = 0$.

Suppose that A has no reciprocal. Then, by the theorems of Chapter 3, the least equation satisfied by A is of the form

$$A^k + c_{k-1}A^{k-1} + \cdots + c_1A = 0.$$

This may be written as

$$A(A^{k-1} + c_{k-1}A^{k-2} + \cdots + c_1 I) = 0.$$

Since A satisfies no equation of degree less than k, if we put

$$B = (A^{k-1} + c_{k-1}A^{k-2} + \cdots + c_1 I),$$

we have $B \neq 0$ and $AB \neq 0$.

If B were a vector, we would be through. The trouble is that we do not know that all the entries in the second, third, etc., columns of B are zero. To remedy this, we use the following trick. Let S be the $n \times n$ matrix

$$\begin{bmatrix} 0 & 0 & 0 & \cdots & 0 \\ 1 & 0 & 0 & \cdots & 0 \\ 0 & 1 & 0 & \cdots & 0 \\ 0 & 0 & 1 & \cdots & 0 \\ \cdot & \cdot & \cdot & \cdot & \cdot \\ 0 & \cdots & 0 & 1 & 0 \end{bmatrix}.$$

It is clear by computation that, if we multiply any matrix C on the right by S, the first column of C disappears, the second, third, etc., columns are all shifted one column to the left, and the final column of the product matrix is simply a column of zeros.

Now, if B is a vector, we are through; if B is not a vector, B has at least one nonzero entry in a column other than the first. This same entry occurs, shifted over one column, in BS; thus $BS \neq 0$. If BS is a vector, then since

$$A(BS) = (AB)S = 0S = 0,$$

we have exhibited a nonzero vector V such that $AV = 0$. If BS is not a vector, then for the same reason as before,

$$(BS)S = BS^2$$

is nonzero. It is perfectly plain that each time we multiply by S, we introduce an extra column of zeros on the right of the resulting matrix. Hence, by multiplying over and over by the matrix S, we must eventually come to a nonzero *vector* of the form

BS^k. Letting $BS^k = V$, we have

$$AV = (AB)S^k = 0 \cdot S^k = 0,$$

while $V \neq 0$, so that our theorem is proved.

EXERCISE

Find a nonzero vector V such that $AV = 0$, A being the matrix

$$\begin{bmatrix} 1 & 0 & 1 & 1 \\ 0 & 1 & 1 & 2 \\ -1 & 0 & 3 & 4 \\ -2 & 2 & 10 & 14 \end{bmatrix}.$$

7-2. Eigenvalues

Let A be a matrix. Then a number x for which $A - xI$ has no reciprocal is called an *eigenvalue* of the matrix A. This notion is fundamental in atomic physics: for the energy levels of atoms and molecules turn out to be given by the eigenvalues of certain matrices.

In the present section, we shall present some of the basic theorems about the eigenvalues of a matrix A. The first theorem is a very easy consequence of the theorem proved in the preceding section.

THEOREM

Let A be a matrix and x a number. Then x is an eigenvalue of A if and only if there exists a nonzero vector V such that

$$AV = xV.$$

The equation $AV = xV$ is called the eigenvalue equation in atomic physics.

PROOF

If x is an eigenvalue of A, so that $A - xI$ has no reciprocal, then by the theorem of the preceding section there exists a nonzero vector V such that $(A - xI)V = 0$, i.e., so that $AV = xV$.

Conversely, suppose that there exists a nonzero vector V such that $AV = xV$. Then $(A - xI)V = 0$, so that by the theorem of the preceding section $A - xI$ has no reciprocal, and x is an eigenvalue of A. Q.E.D.

The importance of eigenvalues in atomic physics makes it important to find ways of calculating them. The following theorem provides one way of doing so.

THEOREM

Let A be a matrix, and let

$$A^k + c_{k-1}A^{k-1} + \cdots + c_0I = 0 \tag{1}$$

be the least equation satisfied by A. Then a number x is an eigenvalue of A if and only if x satisfies the equation

$$x^k + c_{k-1}x^{k-1} + \cdots + c_0 = 0.$$

PROOF

The matrix $B = A - xI$ obviously satisfies the polynomial equation

$$(B + xI)^k + c_{k-1}(B + xI)^{k-1} + \cdots + c_0I = 0. \tag{2}$$

(This equation is not in the standard form in which we have been writing polynomial equations, but it can easily be put in this form by expanding and rearranging according to descending powers of B.)

This equation must be the least equation satisfied by B, since if B satisfied an equation

$$B^j + d_{j-1}B^{j-1} + \cdots + d_0I = 0$$

of a degree j lower than k, then $A = B + xI$ would satisfy the equation

$$(A - xI)^j + d_{j-1}(A - xI)^{j-1} + \cdots + d_0I = 0$$

of degree j less than k, contradicting the fact that (1) is the least equation satisfied by A. Thus (2) is the least equation satisfied by B.

If we put equation (2) in the standard form in which we have been writing polynomial equations by expanding and rearranging according to descending powers of B, we find the lowest term to be

$$(x^k + c_{k-1}x^{k-1} + \cdots + c_0)I.$$

Thus, the last coefficient in the least equation satisfied by B is $x^k + c_{k-1}x^{k-1} + \cdots + c_0$. By the theorem of Chapter 3, B, that is, $A - xI$, fails to have an inverse if and only if this coefficient is zero. Thus our theorem is proved. Q.E.D.

EXERCISES

1. Find the eigenvalues of the matrix

$$A = \begin{bmatrix} 1 & 2 & 3 \\ 2 & 0 & 4 \\ 3 & 4 & 3 \end{bmatrix}$$

accurate to two significant figures.

2. Find the eigenvalues of the matrix (with complex entries)

$$A = \begin{bmatrix} 2 & i \\ -i & 1 \end{bmatrix}.$$

3. A nonzero vector V satisfying $AV = xV$ is called an eigenvector of A (belonging to the eigenvalue x). Find the eigenvectors of the matrix A of Exercise 2.

4. Find all the real and all the complex eigenvalues of the matrix

$$A = \begin{bmatrix} 1 & 2 & 0 & 0 \\ 2 & 3 & 0 & 0 \\ 0 & 0 & 3 & 4 \\ 0 & 0 & 4 & 5 \end{bmatrix}.$$

(The teacher may assign additional exercises of this type.)

Chapter 8

INFINITE SERIES OF MATRICES

8-1. The Geometric Series

We have observed repeatedly in earlier sections that the algebra of matrices has many similarities with ordinary algebra and that this fact can be used to carry over all sorts of principles and theorems from ordinary algebra to vector algebra. Thus, algebraic results which we know from elementary algebra are seen to apply in a vastly more general setting, take on new richness and scope, permit many new applications, etc. In the present theorem, we will give two examples designed to show how even some of the more subtle results of algebra can be applied to matrices.

In ordinary algebra we sum the geometric series

$$1 + x + \cdots + x^n = \frac{1}{1-x} - \frac{1}{1-x}\, x^{n+1}.$$

If the number x is less than 1 in absolute value, then multiplying it by itself gives a sequence of numbers whose absolute values rapidly become exceedingly small. It is then possible to take the "limit" of the above expression and write the "infinite series"

$$1 + x + \cdots + x^n + \cdots = \frac{1}{1-x}. \tag{1}$$

This procedure can actually be justified perfectly rigorously in terms of the formal notions of limits and convergence introduced in higher mathematics, but we shall not attempt to do so in this book.

Nevertheless, the formula is extremely suggestive and leads us to imagine that for matrices we have a corresponding formula

$$I + A + A^2 + A^3 + \cdots = (I - A)^{-1}. \tag{2}$$

Formula (1) holds only if the absolute value of x is sufficiently small, so that it is only reasonable to expect that formula (2) will hold if the entries in the matrix A are sufficiently small; this point will be discussed more carefully in the next section. In the present section, we will simply assume that the "infinite sum" $I + A + A^2 + A^3 + \cdots$ makes sense and that it can be manipulated by the algebraic laws governing finite sums. If this is the case, we can argue as follows:

Let $B = I + A + A^2 + \cdots$. Then

$$\begin{aligned}(I - A)B &= B - AB\\ &= (I + A + A^2 + \cdots) - A(I + A + A^2 + \cdots)\\ &= (I + A + A^2 + \cdots) - (A + A^2 + A^3 + \cdots)\\ &= I.\end{aligned}$$

Thus $(I - A)B = I$, so that B is the reciprocal of $I - A$, and formula (2) is proved.

Formula (2) is often extremely useful in making approximate calculations of the inverses of large matrices.

Suppose that we want the inverse of the 5×5 matrix

$$B = \begin{bmatrix} 0.99 & 0 & 0.01 & 0 & 0.03 \\ 0 & 1.01 & 0 & -0.01 & 0.04 \\ 0.02 & 0 & 0.98 & 0 & 0 \\ 0 & 0.01 & 0.01 & 1.02 & -0.01 \\ 0.01 & 0 & -0.01 & 0 & 0.97 \end{bmatrix}.$$

This matrix may be written as $I - A$, where A is the matrix

$$A = \begin{bmatrix} 0.01 & 0 & -0.01 & 0 & -0.03 \\ 0 & -0.01 & 0 & 0.01 & -0.04 \\ -0.02 & 0 & 0.02 & 0 & 0 \\ 0 & -0.01 & -0.01 & -0.02 & 0.01 \\ -0.01 & 0 & 0.01 & 0 & 0.03 \end{bmatrix}.$$

Hence to two significant figures, the reciprocal of B is

$$B^{-1} \cong I + A = \begin{bmatrix} 1.01 & 0 & -0.01 & 0 & -0.03 \\ 0 & 0.99 & 0 & 0.01 & -0.04 \\ -0.02 & 0 & 1.02 & 0 & 0 \\ 0 & -0.01 & -0.01 & 0.98 & -0.01 \\ -0.01 & 0 & 0.01 & 0 & 1.03 \end{bmatrix}.$$

EXERCISES

1. Calculate the inverse of the matrix B of the preceding section to four significant figures.

2. Let B be the 5×5 matrix

$$B = \begin{bmatrix} 1.01 & 0 & -0.01 & 0 & -0.03 \\ 0 & 1.99 & 0 & 0.01 & -0.04 \\ -0.02 & 0 & 3.02 & 0 & 0 \\ 0 & -0.01 & -0.01 & 3.98 & 0.01 \\ -0.01 & 0 & 0.01 & 0 & 5.03 \end{bmatrix}.$$

Calculate the reciprocal of B, accurate to four significant figures. [*Hint:* Write $B = CD$, where

$$C = \begin{bmatrix} 1 & 0 & 0 & 0 & 0 \\ 0 & 2 & 0 & 0 & 0 \\ 0 & 0 & 3 & 0 & 0 \\ 0 & 0 & 0 & 4 & 0 \\ 0 & 0 & 0 & 0 & 5 \end{bmatrix},$$

and use the law $D^{-1}C^{-1} = (CD)^{-1}$.]

3. Solve the system of equations

$$\begin{aligned} 1.02x_1 - 0.01x_3 - 0.03x_5 &= 1 \\ 1.98x_2 + 0.01x_4 - 0.04x_5 &= 0 \\ 0.02x_1 + 3.03x_3 &= 0 \\ -0.01x_2 - 0.01x_3 + 3.98x_4 + 0.01x_5 &= 1 \\ -0.01x_1 + 0.01x_3 + 5.04x_5 &= 0, \end{aligned}$$

getting an answer accurate to four significant figures.

8-2. The Size of the Entries in the Power of A

For the infinite sum $I + A + A^2 + A^3 + \cdots$ in the preceding section to make sense, it is necessary (just as in $1 + x + x^2 + \cdots$) that the entries in the successive powers of A grow smaller and smaller. Thus, to complete the discussion of the preceding section, we need to have an estimate for the size of the entries in the powers of A.

Such an estimate is given in the following theorem.

THEOREM

Let A be an $n \times n$ matrix and c a positive number. Suppose that every entry in A has an absolute value less than c. Then every entry in A^k has an absolute value less than

$$(nc)^k.$$

PROOF

We will get the result by using mathematical induction on k. If $k = 1$, then since $nc \geq c$, the theorem is true. Suppose it to be true for some value k_0 of k. Then the entry in the ith row and jth column of A^{k_0+1} is

$$[A^{k_0+1}]_{i,j} = [A^{k_0}]_{i,1}[A]_{1,j} + [A^{k_0}]_{i,2}[A]_{2,j} + \cdots + [A^{k_0}]_{i,n}[A]_{n,j}. \tag{3}$$

Now, if we have a sum of n numbers $x_1, \ldots, x_n$, we have

$x_1 \leq |x_1|, \ldots, x_n \leq |x_n|$, and adding all these inequalities, we have

$$x_1 + \cdots + x_n \leq |x_1| + \cdots + |x_n|.$$

Similarly, $-x_1 \leq |x_1|$, $-x_2 \leq |x_2|, \ldots, -x_n \leq |x_n|$, so that adding these inequalities, we get

$$-(x_1 + \cdots + x_n) \leq |x_1| + \cdots + |x_n|.$$

Since the absolute value of a quantity is simply the larger of the quantity itself and of its negative, it follows that

$$|x_1 + \cdots + x_n| \leq |x_1| + \cdots + |x_n|.$$

Applying this general inequality to formula (1) and using the identity $|xy| = |x|\,|y|$, we find

$$|[A^{k_0+1}]_{i,j}| \leq |[A^{k_0}]_{i,1}|\,|[A]_{1,j}| + |[A^{k_0}]_{i,2}|\,|[A]_{2,j}| + \cdots + |[A^{k_0}]_{i,n}|\,|[A]_{n,j}|.$$

Since by hypothesis the absolute value of every entry of A is less than c, and since by inductive assumption the absolute value of every entry of A^{k_0} is less than $(cn)^{k_0}$, we find at once that

$$|A^{k_0+1}|_{i,j} < \underbrace{(cn)^{k_0} \cdot c + (cn)^{k_0} \cdot c + \cdots + (cn)^{k_0} \cdot c}_{n - \text{times}}$$
$$= n \cdot c \cdot (cn)^{k_0} = (cn)^{k_0+1}.$$

Thus if our theorem is true for $k = k_0$, it is true for $k = k_0 + 1$. This completes our inductive step, and the theorem now follows by the principle of mathematical induction. Q.E.D.

We may apply this theorem to the situation studied in the preceding section as follows: if d is a positive number less than 1, and if every entry in the $n \times n$ matrix A has an absolute value less than the quotient d/n, then it follows by our theorem that every element in any power A^k is less than d^k. The numbers in the sequence d, d^2, d^3, d^4 quite rapidly become exceedingly small; hence the entries in the matrix A^k must be extremely small if k is large.

8-3. The Exponential Series

A very important function of the number x is the exponential function e^x, e being the base of natural logarithms. As you know, this function satisfies the law of exponents

$$e^x e^y = e^{x+y}. \tag{4}$$

As you also know, it is given by the infinite sum

$$e^x = 1 + \frac{x}{1!} + \frac{x^2}{2!} + \cdots .$$

In accordance with our general principle of trying to carry over even these deeper facts about numbers to matrices, we define the exponential matrix of a matrix A to be the value of the infinite sum of matrices

$$e^A = I + \frac{1}{1!}A + \frac{1}{2!}A^2 + \frac{1}{3!}A^3 + \cdots .$$

It would be nice if the equation $e^A e^B = e^{A+B}$, generalizing the equation (4), held for all matrices A and B. Unfortunately, this formula is not always true. How could it be? Because of the failure of the commutative law of multiplication, we have to expect that we can have $e^A e^B \neq e^B e^A$. On the other hand, since $A + B = B + A$ is always true, we must have $e^{A+B} = e^{B+A}$.

Nevertheless, we shall see that, with the additional hypothesis that $AB = BA$, the desired formula does hold. As a preliminary, we make the following definition.

DEFINITION

Let A and B be two matrices of the same size. Then A and B are said to *commute with* each other, if $AB = BA$.

If two matrices commute with each other, then algebraic expressions involving just these two matrices, and no other, can be multiplied, added, etc., just as if they were expressions in the ordinary algebra of numbers. The following theorem notes one particular aspect of this general principle.

THEOREM

Let A and B commute with each other. Then every power of A commutes with every power of B.

PROOF

We shall show that A^2 commutes with B^3; the general proof may be given by using mathematical induction, and is left to the student as an exercise.

Using the fact that $AB = BA$, we have

$$
\begin{aligned}
A^2B^3 = AABBB = ABABB = ABBAB & \\
&= ABBBA \\
&= BABBA = BBABA = BBBAA \\
&= B^3A^2.
\end{aligned}
$$

Q.E.D.

Since all powers of A commute with all powers of B, it follows, as remarked above, that algebraic expressions with numerical coefficients involving just these two matrices, and no others, can be multiplied, added, etc., just as if they were expressions in the ordinary algebra of numbers. Thus the binomial expansion for the product $(A + B)^n$ must be valid. We state this as a formal theorem, expressing the theorem in Σ notation.

THEOREM

Let A and B be two matrices of the same size which commute with each other. Then

$$(A + B)^k = \sum_{j=0}^{k} \frac{k!}{j!\,k - j!} A^jB^{k-j}.$$

The proof by mathematical induction, which is formally indistinguishable from the corresponding proof in the algebra of real numbers, is left to the student as an exercise.

If now A and B are two matrices of the same size commuting with each other, and we multiply out the two sums defining e^A and e^B and collect all terms which have a given sum of exponents, we find

$$e^Ae^B = \left(I + \frac{A}{1!} + \frac{A^2}{2!} + \cdots\right)\left(I + \frac{B}{1!} + \frac{B^2}{2!} + \cdots\right)$$

$$e^Ae^B = I + \left(\frac{A}{1!}\frac{B}{1!}\right) + \left(\frac{A^2}{2!} + \frac{A}{1!}\frac{B}{1!} + \frac{B^2}{2!}\right) + \cdots$$
$$+ \left(\frac{A^n}{n!} + \frac{A^{n-1}}{n-1!}\frac{B}{1!} + \frac{A^{n-2}}{n-2!}\frac{B}{2!} + \cdots\right.$$
$$\left.+ \frac{A^2}{2!}\frac{B^{n-2}}{n-2!} + \frac{A}{1!}\frac{B^{n-1}}{n-1!}\frac{B^n}{n!}\right) + \cdots$$

$$e^Ae^B = I + \frac{1}{1!}(A + B) + \frac{1}{2!}(A^2 + 2AB + B^2) + \cdots$$
$$+ \frac{1}{n!}\left(\sum_{j=0}^{n} \frac{n!}{j!n-j!} A^jB^{n-j}\right) + \cdots$$

so that using the binomial theorem

$$e^Ae^B = I + \frac{1}{1!}(A + B) + \frac{1}{2!}(A + B)^2 + \cdots + \frac{1}{n!}(A + B)^n$$
$$+ \cdots = e^{A+B}$$

and we have proved the desired formula. Q.E.D.

EXERCISES

1. Show that if a 4×4 matrix A commutes with the matrix

$$\begin{bmatrix} 1 & 0 & 0 & 0 \\ 0 & 2 & 0 & 0 \\ 0 & 0 & 3 & 0 \\ 0 & 0 & 0 & 4 \end{bmatrix}$$

then $[A]_{i,j} = 0$ unless $i = j$.

2. Prove that the only 4×4 matrices A which commute with every other 4×4 matrix B are the numerical multiples of the identity matrix.

3. If

$$A = \begin{bmatrix} x & 0 & 0 & 0 \\ 0 & y & 0 & 0 \\ 0 & 0 & z & 0 \\ 0 & 0 & 0 & t \end{bmatrix}$$

show that e^A is the matrix

$$\begin{bmatrix} e^x & 0 & 0 & 0 \\ 0 & e^y & 0 & 0 \\ 0 & 0 & e^z & 0 \\ 0 & 0 & 0 & e^t \end{bmatrix}.$$

4. If

$$A = \begin{bmatrix} x & y \\ -y & x \end{bmatrix}$$

show that

$$\begin{bmatrix} e^x \cos y & e^x \sin y \\ -e^x \sin y & e^x \cos y \end{bmatrix}.$$

(*Hint:* Use Euler's formula $e^{iy} = \cos y + i \sin y$.)

5. Prove the first theorem of the preceding section.

6. Prove the second theorem of the preceding section.

INDEX

A CATALOG OF SELECTED

DOVER BOOKS

IN ALL FIELDS OF INTEREST

A CATALOG OF SELECTED DOVER BOOKS IN ALL FIELDS OF INTEREST

CONCERNING THE SPIRITUAL IN ART, Wassily Kandinsky. Pioneering work by father of abstract art. Thoughts on color theory, nature of art. Analysis of earlier masters. 12 illustrations. 80pp. of text. 5⅜ x 8½. 23411-8 Pa. $4.95

ANIMALS: 1,419 Copyright-Free Illustrations of Mammals, Birds, Fish, Insects, etc., Jim Harter (ed.). Clear wood engravings present, in extremely lifelike poses, over 1,000 species of animals. One of the most extensive pictorial sourcebooks of its kind. Captions. Index. 284pp. 9 x 12. 23766-4 Pa. $14.95

CELTIC ART: The Methods of Construction, George Bain. Simple geometric techniques for making Celtic interlacements, spirals, Kells-type initials, animals, humans, etc. Over 500 illustrations. 160pp. 9 x 12. (Available in U.S. only.) 22923-8 Pa. $9.95

AN ATLAS OF ANATOMY FOR ARTISTS, Fritz Schider. Most thorough reference work on art anatomy in the world. Hundreds of illustrations, including selections from works by Vesalius, Leonardo, Goya, Ingres, Michelangelo, others. 593 illustrations. 192pp. 7⅛ x 10¼. 20241-0 Pa. $9.95

CELTIC HAND STROKE-BY-STROKE (Irish Half-Uncial from "The Book of Kells"): An Arthur Baker Calligraphy Manual, Arthur Baker. Complete guide to creating each letter of the alphabet in distinctive Celtic manner. Covers hand position, strokes, pens, inks, paper, more. Illustrated. 48pp. 8¼ x 11. 24336-2 Pa. $3.95

EASY ORIGAMI, John Montroll. Charming collection of 32 projects (hat, cup, pelican, piano, swan, many more) specially designed for the novice origami hobbyist. Clearly illustrated easy-to-follow instructions insure that even beginning papercrafters will achieve successful results. 48pp. 8¼ x 11. 27298-2 Pa. $3.50

THE COMPLETE BOOK OF BIRDHOUSE CONSTRUCTION FOR WOODWORKERS, Scott D. Campbell. Detailed instructions, illustrations, tables. Also data on bird habitat and instinct patterns. Bibliography. 3 tables. 63 illustrations in 15 figures. 48pp. 5¼ x 8½. 24407-5 Pa. $2.50

BLOOMINGDALE'S ILLUSTRATED 1886 CATALOG: Fashions, Dry Goods and Housewares, Bloomingdale Brothers. Famed merchants' extremely rare catalog depicting about 1,700 products: clothing, housewares, firearms, dry goods, jewelry, more. Invaluable for dating, identifying vintage items. Also, copyright-free graphics for artists, designers. Co-published with Henry Ford Museum & Greenfield Village. 160pp. 8¼ x 11. 25780-0 Pa. $12.95

HISTORIC COSTUME IN PICTURES, Braun & Schneider. Over 1,450 costumed figures in clearly detailed engravings–from dawn of civilization to end of 19th century. Captions. Many folk costumes. 256pp. 8⅜ x 11¾. 23150-X Pa. $12.95

THE CLARINET AND CLARINET PLAYING, David Pino. Lively, comprehensive work features suggestions about technique, musicianship, and musical interpretation, as well as guidelines for teaching, making your own reeds, and preparing for public performance. Includes an intriguing look at clarinet history. "A godsend," *The Clarinet,* Journal of the International Clarinet Society. Appendixes. 7 illus. 320pp. 5⅜ x 8½. 40270-3 Pa. $9.95

HOLLYWOOD GLAMOR PORTRAITS, John Kobal (ed.). 145 photos from 1926-49. Harlow, Gable, Bogart, Bacall; 94 stars in all. Full background on photographers, technical aspects. 160pp. 8⅜ x 11¼. 23352-9 Pa. $12.95

THE ANNOTATED CASEY AT THE BAT: A Collection of Ballads about the Mighty Casey/Third, Revised Edition, Martin Gardner (ed.). Amusing sequels and parodies of one of America's best-loved poems: Casey's Revenge, Why Casey Whiffed, Casey's Sister at the Bat, others. 256pp. 5⅜ x 8½. 28598-7 Pa. $8.95

THE RAVEN AND OTHER FAVORITE POEMS, Edgar Allan Poe. Over 40 of the author's most memorable poems: "The Bells," "Ulalume," "Israfel," "To Helen," "The Conqueror Worm," "Eldorado," "Annabel Lee," many more. Alphabetic lists of titles and first lines. 64pp. 5$\frac{3}{16}$ x 8¼. 26685-0 Pa. $1.00

PERSONAL MEMOIRS OF U. S. GRANT, Ulysses Simpson Grant. Intelligent, deeply moving firsthand account of Civil War campaigns, considered by many the finest military memoirs ever written. Includes letters, historic photographs, maps and more. 528pp. 6⅛ x 9¼. 28587-1 Pa. $12.95

ANCIENT EGYPTIAN MATERIALS AND INDUSTRIES, A. Lucas and J. Harris. Fascinating, comprehensive, thoroughly documented text describes this ancient civilization's vast resources and the processes that incorporated them in daily life, including the use of animal products, building materials, cosmetics, perfumes and incense, fibers, glazed ware, glass and its manufacture, materials used in the mummification process, and much more. 544pp. 6⅛ x 9¼. (Available in U.S. only.) 40446-3 Pa. $16.95

RUSSIAN STORIES/PYCCKNE PACCKA3bl: A Dual-Language Book, edited by Gleb Struve. Twelve tales by such masters as Chekhov, Tolstoy, Dostoevsky, Pushkin, others. Excellent word-for-word English translations on facing pages, plus teaching and study aids, Russian/English vocabulary, biographical/critical introductions, more. 416pp. 5⅜ x 8½. 26244-8 Pa. $9.95

PHILADELPHIA THEN AND NOW: 60 Sites Photographed in the Past and Present, Kenneth Finkel and Susan Oyama. Rare photographs of City Hall, Logan Square, Independence Hall, Betsy Ross House, other landmarks juxtaposed with contemporary views. Captures changing face of historic city. Introduction. Captions. 128pp. 8¼ x 11. 25790-8 Pa. $9.95

AIA ARCHITECTURAL GUIDE TO NASSAU AND SUFFOLK COUNTIES, LONG ISLAND, The American Institute of Architects, Long Island Chapter, and the Society for the Preservation of Long Island Antiquities. Comprehensive, well-researched and generously illustrated volume brings to life over three centuries of Long Island's great architectural heritage. More than 240 photographs with authoritative, extensively detailed captions. 176pp. 8¼ x 11. 26946-9 Pa. $14.95

NORTH AMERICAN INDIAN LIFE: Customs and Traditions of 23 Tribes, Elsie Clews Parsons (ed.). 27 fictionalized essays by noted anthropologists examine religion, customs, government, additional facets of life among the Winnebago, Crow, Zuni, Eskimo, other tribes. 480pp. 6⅛ x 9¼. 27377-6 Pa. $10.95

FRANK LLOYD WRIGHT'S DANA HOUSE, Donald Hoffmann. Pictorial essay of residential masterpiece with over 160 interior and exterior photos, plans, elevations, sketches and studies. 128pp. $9^{1}/_{4}$ x $10^{3}/_{4}$. 29120-0 Pa. $14.95

THE MALE AND FEMALE FIGURE IN MOTION: 60 Classic Photographic Sequences, Eadweard Muybridge. 60 true-action photographs of men and women walking, running, climbing, bending, turning, etc., reproduced from rare 19th-century masterpiece. vi + 121pp. 9 x 12. 24745-7 Pa. $12.95

1001 QUESTIONS ANSWERED ABOUT THE SEASHORE, N. J. Berrill and Jacquelyn Berrill. Queries answered about dolphins, sea snails, sponges, starfish, fishes, shore birds, many others. Covers appearance, breeding, growth, feeding, much more. 305pp. 5¼ x 8¼. 23366-9 Pa. $9.95

ATTRACTING BIRDS TO YOUR YARD, William J. Weber. Easy-to-follow guide offers advice on how to attract the greatest diversity of birds: birdhouses, feeders, water and waterers, much more. 96pp. $5^{3}/_{16}$ x $8^{1}/_{4}$. 28927-3 Pa. $2.50

MEDICINAL AND OTHER USES OF NORTH AMERICAN PLANTS: A Historical Survey with Special Reference to the Eastern Indian Tribes, Charlotte Erichsen-Brown. Chronological historical citations document 500 years of usage of plants, trees, shrubs native to eastern Canada, northeastern U.S. Also complete identifying information. 343 illustrations. 544pp. 6½ x 9¼. 25951-X Pa. $12.95

STORYBOOK MAZES, Dave Phillips. 23 stories and mazes on two-page spreads: Wizard of Oz, Treasure Island, Robin Hood, etc. Solutions. 64pp. 8¼ x 11. 23628-5 Pa. $2.95

AMERICAN NEGRO SONGS: 230 Folk Songs and Spirituals, Religious and Secular, John W. Work. This authoritative study traces the African influences of songs sung and played by black Americans at work, in church, and as entertainment. The author discusses the lyric significance of such songs as "Swing Low, Sweet Chariot," "John Henry," and others and offers the words and music for 230 songs. Bibliography. Index of Song Titles. 272pp. $6^{1}/_{2}$ x $9^{1}/_{4}$. 40271-1 Pa. $10.95

MOVIE-STAR PORTRAITS OF THE FORTIES, John Kobal (ed.). 163 glamor, studio photos of 106 stars of the 1940s: Rita Hayworth, Ava Gardner, Marlon Brando, Clark Gable, many more. 176pp. 8⅜ x 11¼. 23546-7 Pa. $14.95

BENCHLEY LOST AND FOUND, Robert Benchley. Finest humor from early 30s, about pet peeves, child psychologists, post office and others. Mostly unavailable elsewhere. 73 illustrations by Peter Arno and others. 183pp. 5⅜ x 8½. 22410-4 Pa. $6.95

YEKL and THE IMPORTED BRIDEGROOM AND OTHER STORIES OF YIDDISH NEW YORK, Abraham Cahan. Film Hester Street based on *Yekl* (1896). Novel, other stories among first about Jewish immigrants on N.Y.'s East Side. 240pp. 5⅜ x 8½. 22427-9 Pa. $7.95

SELECTED POEMS, Walt Whitman. Generous sampling from *Leaves of Grass*. Twenty-four poems include "I Hear America Singing," "Song of the Open Road," "I Sing the Body Electric," "When Lilacs Last in the Dooryard Bloom'd," "O Captain! My Captain!"–all reprinted from an authoritative edition. Lists of titles and first lines. 128pp. $5^{3}/_{16}$ x 8¼. 26878-0 Pa. $1.00

MY BONDAGE AND MY FREEDOM, Frederick Douglass. Born a slave, Douglass became outspoken force in antislavery movement. The best of Douglass' autobiographies. Graphic description of slave life. 464pp. 5⅜ x 8½. 22457-0 Pa. $8.95

FOLLOWING THE EQUATOR: A Journey Around the World, Mark Twain. Fascinating humorous account of 1897 voyage to Hawaii, Australia, India, New Zealand, etc. Ironic, bemused reports on peoples, customs, climate, flora and fauna, politics, much more. 197 illustrations. 720pp. 5⅜ x 8½. 26113-1 Pa. $15.95

THE PEOPLE CALLED SHAKERS, Edward D. Andrews. Definitive study of Shakers: origins, beliefs, practices, dances, social organization, furniture and crafts, etc. 33 illustrations. 351pp. 5⅜ x 8½. 21081-2 Pa. $12.95

THE MYTHS OF GREECE AND ROME, H. A. Guerber. A classic of mythology, generously illustrated, long prized for its simple, graphic, accurate retelling of the principal myths of Greece and Rome, and for its commentary on their origins and significance. With 64 illustrations by Michelangelo, Raphael, Titian, Rubens, Canova, Bernini and others. 480pp. 5⅜ x 8½. 27584-1 Pa. $10.95

PSYCHOLOGY OF MUSIC, Carl E. Seashore. Classic work discusses music as a medium from psychological viewpoint. Clear treatment of physical acoustics, auditory apparatus, sound perception, development of musical skills, nature of musical feeling, host of other topics. 88 figures. 408pp. 5⅜ x 8½. 21851-1 Pa. $11.95

THE PHILOSOPHY OF HISTORY, Georg W. Hegel. Great classic of Western thought develops concept that history is not chance but rational process, the evolution of freedom. 457pp. 5⅜ x 8½. 20112-0 Pa. $9.95

THE BOOK OF TEA, Kakuzo Okakura. Minor classic of the Orient: entertaining, charming explanation, interpretation of traditional Japanese culture in terms of tea ceremony. 94pp. 5⅜ x 8½. 20070-1 Pa. $3.95

LIFE IN ANCIENT EGYPT, Adolf Erman. Fullest, most thorough, detailed older account with much not in more recent books, domestic life, religion, magic, medicine, commerce, much more. Many illustrations reproduce tomb paintings, carvings, hieroglyphs, etc. 597pp. 5⅜ x 8½. 22632-8 Pa. $12.95

SUNDIALS, Their Theory and Construction, Albert Waugh. Far and away the best, most thorough coverage of ideas, mathematics concerned, types, construction, adjusting anywhere. Simple, nontechnical treatment allows even children to build several of these dials. Over 100 illustrations. 230pp. 5⅜ x 8½. 22947-5 Pa. $8.95

THEORETICAL HYDRODYNAMICS, L. M. Milne-Thomson. Classic exposition of the mathematical theory of fluid motion, applicable to both hydrodynamics and aerodynamics. Over 600 exercises. 768pp. 6⅛ x 9¼. 68970-0 Pa. $20.95

SONGS OF EXPERIENCE: Facsimile Reproduction with 26 Plates in Full Color, William Blake. 26 full-color plates from a rare 1826 edition. Includes "The Tyger," "London," "Holy Thursday," and other poems. Printed text of poems. 48pp. 5¼ x 7. 24636-1 Pa. $4.95

OLD-TIME VIGNETTES IN FULL COLOR, Carol Belanger Grafton (ed.). Over 390 charming, often sentimental illustrations, selected from archives of Victorian graphics–pretty women posing, children playing, food, flowers, kittens and puppies, smiling cherubs, birds and butterflies, much more. All copyright-free. 48pp. 9¼ x 12¼. 27269-9 Pa. $9.95

PERSPECTIVE FOR ARTISTS, Rex Vicat Cole. Depth, perspective of sky and sea, shadows, much more, not usually covered. 391 diagrams, 81 reproductions of drawings and paintings. 279pp. 5⅜ x 8½. 22487-2 Pa. $9.95

DRAWING THE LIVING FIGURE, Joseph Sheppard. Innovative approach to artistic anatomy focuses on specifics of surface anatomy, rather than muscles and bones. Over 170 drawings of live models in front, back and side views, and in widely varying poses. Accompanying diagrams. 177 illustrations. Introduction. Index. 144pp. 8⅜ x11¼. 26723-7 Pa. $9.95

GOTHIC AND OLD ENGLISH ALPHABETS: 100 Complete Fonts, Dan X. Solo. Add power, elegance to posters, signs, other graphics with 100 stunning copyright-free alphabets: Blackstone, Dolbey, Germania, 97 more–including many lower-case, numerals, punctuation marks. 104pp. 8⅛ x 11. 24695-7 Pa. $9.95

HOW TO DO BEADWORK, Mary White. Fundamental book on craft from simple projects to five-bead chains and woven works. 106 illustrations. 142pp. 5⅜ x 8. 20697-1 Pa. $5.95

THE BOOK OF WOOD CARVING, Charles Marshall Sayers. Finest book for beginners discusses fundamentals and offers 34 designs. "Absolutely first rate . . . well thought out and well executed."–E. J. Tangerman. 118pp. 7¾ x 10⅝. 23654-4 Pa. $7.95

ILLUSTRATED CATALOG OF CIVIL WAR MILITARY GOODS: Union Army Weapons, Insignia, Uniform Accessories, and Other Equipment, Schuyler, Hartley, and Graham. Rare, profusely illustrated 1846 catalog includes Union Army uniform and dress regulations, arms and ammunition, coats, insignia, flags, swords, rifles, etc. 226 illustrations. 160pp. 9 x 12. 24939-5 Pa. $12.95

WOMEN'S FASHIONS OF THE EARLY 1900s: An Unabridged Republication of "New York Fashions, 1909," National Cloak & Suit Co. Rare catalog of mail-order fashions documents women's and children's clothing styles shortly after the turn of the century. Captions offer full descriptions, prices. Invaluable resource for fashion, costume historians. Approximately 725 illustrations. 128pp. 8⅜ x 11¼. 27276-1 Pa. $12.95

THE 1912 AND 1915 GUSTAV STICKLEY FURNITURE CATALOGS, Gustav Stickley. With over 200 detailed illustrations and descriptions, these two catalogs are essential reading and reference materials and identification guides for Stickley furniture. Captions cite materials, dimensions and prices. 112pp. 6½ x 9¼. 26676-1 Pa. $9.95

EARLY AMERICAN LOCOMOTIVES, John H. White, Jr. Finest locomotive engravings from early 19th century: historical (1804–74), main-line (after 1870), special, foreign, etc. 147 plates. 142pp. 11⅜ x 8¼. 22772-3 Pa. $12.95

THE TALL SHIPS OF TODAY IN PHOTOGRAPHS, Frank O. Braynard. Lavishly illustrated tribute to nearly 100 majestic contemporary sailing vessels: Amerigo Vespucci, Clearwater, Constitution, Eagle, Mayflower, Sea Cloud, Victory, many more. Authoritative captions provide statistics, background on each ship. 190 black-and-white photographs and illustrations. Introduction. 128pp. 8⅞ x 11¾. 27163-3 Pa. $14.95

LITTLE BOOK OF EARLY AMERICAN CRAFTS AND TRADES, Peter Stockham (ed.). 1807 children's book explains crafts and trades: baker, hatter, cooper, potter, and many others. 23 copperplate illustrations. 140pp. 4⅝ x 6.
23336-7 Pa. $4.95

VICTORIAN FASHIONS AND COSTUMES FROM HARPER'S BAZAR, 1867–1898, Stella Blum (ed.). Day costumes, evening wear, sports clothes, shoes, hats, other accessories in over 1,000 detailed engravings. 320pp. 9⅜ x 12¼.
22990-4 Pa. $16.95

GUSTAV STICKLEY, THE CRAFTSMAN, Mary Ann Smith. Superb study surveys broad scope of Stickley's achievement, especially in architecture. Design philosophy, rise and fall of the Craftsman empire, descriptions and floor plans for many Craftsman houses, more. 86 black-and-white halftones. 31 line illustrations. Introduction 208pp. 6½ x 9¼. 27210-9 Pa. $9.95

THE LONG ISLAND RAIL ROAD IN EARLY PHOTOGRAPHS, Ron Ziel. Over 220 rare photos, informative text document origin (1844) and development of rail service on Long Island. Vintage views of early trains, locomotives, stations, passengers, crews, much more. Captions. 8⅞ x 11¾. 26301-0 Pa. $14.95

VOYAGE OF THE LIBERDADE, Joshua Slocum. Great 19th-century mariner's thrilling, first-hand account of the wreck of his ship off South America, the 35-foot boat he built from the wreckage, and its remarkable voyage home. 128pp. 5⅜ x 8½.
40022-0 Pa. $5.95

TEN BOOKS ON ARCHITECTURE, Vitruvius. The most important book ever written on architecture. Early Roman aesthetics, technology, classical orders, site selection, all other aspects. Morgan translation. 331pp. 5⅜ x 8½. 20645-9 Pa. $9.95

THE HUMAN FIGURE IN MOTION, Eadweard Muybridge. More than 4,500 stopped-action photos, in action series, showing undraped men, women, children jumping, lying down, throwing, sitting, wrestling, carrying, etc. 390pp. 7⅞ x 10⅝.
20204-6 Clothbd. $29.95

TREES OF THE EASTERN AND CENTRAL UNITED STATES AND CANADA, William M. Harlow. Best one-volume guide to 140 trees. Full descriptions, woodlore, range, etc. Over 600 illustrations. Handy size. 288pp. 4½ x 6⅜.
20395-6 Pa. $6.95

SONGS OF WESTERN BIRDS, Dr. Donald J. Borror. Complete song and call repertoire of 60 western species, including flycatchers, juncoes, cactus wrens, many more–includes fully illustrated booklet. Cassette and manual 99913-0 $8.95

GROWING AND USING HERBS AND SPICES, Milo Miloradovich. Versatile handbook provides all the information needed for cultivation and use of all the herbs and spices available in North America. 4 illustrations. Index. Glossary. 236pp. 5⅜ x 8½.
25058-X Pa. $7.95

BIG BOOK OF MAZES AND LABYRINTHS, Walter Shepherd. 50 mazes and labyrinths in all–classical, solid, ripple, and more–in one great volume. Perfect inexpensive puzzler for clever youngsters. Full solutions. 112pp. 8⅛ x 11.
22951-3 Pa. $5.95

PIANO TUNING, J. Cree Fischer. Clearest, best book for beginner, amateur. Simple repairs, raising dropped notes, tuning by easy method of flattened fifths. No previous skills needed. 4 illustrations. 201pp. 5⅜ x 8½. 23267-0 Pa. $6.95

HINTS TO SINGERS, Lillian Nordica. Selecting the right teacher, developing confidence, overcoming stage fright, and many other important skills receive thoughtful discussion in this indispensible guide, written by a world-famous diva of four decades' experience. 96pp. 5³/₈ x 8¹/₂. 40094-8 Pa. $4.95

THE COMPLETE NONSENSE OF EDWARD LEAR, Edward Lear. All nonsense limericks, zany alphabets, Owl and Pussycat, songs, nonsense botany, etc., illustrated by Lear. Total of 320pp. 5⅜ x 8½. (Available in U.S. only.) 20167-8 Pa. $7.95

VICTORIAN PARLOUR POETRY: An Annotated Anthology, Michael R. Turner. 117 gems by Longfellow, Tennyson, Browning, many lesser-known poets. "The Village Blacksmith," "Curfew Must Not Ring Tonight," "Only a Baby Small," dozens more, often difficult to find elsewhere. Index of poets, titles, first lines. xxiii + 325pp. 5⅝ x 8¼. 27044-0 Pa. $12.95

DUBLINERS, James Joyce. Fifteen stories offer vivid, tightly focused observations of the lives of Dublin's poorer classes. At least one, "The Dead," is considered a masterpiece. Reprinted complete and unabridged from standard edition. 160pp. 5³⁄₁₆ x 8¼. 26870-5 Pa. $1.50

GREAT WEIRD TALES: 14 Stories by Lovecraft, Blackwood, Machen and Others, S. T. Joshi (ed.). 14 spellbinding tales, including "The Sin Eater," by Fiona McLeod, "The Eye Above the Mantel," by Frank Belknap Long, as well as renowned works by R. H. Barlow, Lord Dunsany, Arthur Machen, W. C. Morrow and eight other masters of the genre. 256pp. 5⅜ x 8½. (Available in U.S. only.) 40436-6 Pa. $8.95

THE BOOK OF THE SACRED MAGIC OF ABRAMELIN THE MAGE, translated by S. MacGregor Mathers. Medieval manuscript of ceremonial magic. Basic document in Aleister Crowley, Golden Dawn groups. 268pp. 5⅜ x 8½. 23211-5 Pa. $9.95

NEW RUSSIAN-ENGLISH AND ENGLISH-RUSSIAN DICTIONARY, M. A. O'Brien. This is a remarkably handy Russian dictionary, containing a surprising amount of information, including over 70,000 entries. 366pp. 4½ x 6⅛. 20208-9 Pa. $10.95

HISTORIC HOMES OF THE AMERICAN PRESIDENTS, Second, Revised Edition, Irvin Haas. A traveler's guide to American Presidential homes, most open to the public, depicting and describing homes occupied by every American President from George Washington to George Bush. With visiting hours, admission charges, travel routes. 175 photographs. Index. 160pp. 8¼ x 11. 26751-2 Pa. $13.95

NEW YORK IN THE FORTIES, Andreas Feininger. 162 brilliant photographs by the well-known photographer, formerly with *Life* magazine. Commuters, shoppers, Times Square at night, much else from city at its peak. Captions by John von Hartz. 181pp. 9¼ x 10¾. 23585-8 Pa. $13.95

INDIAN SIGN LANGUAGE, William Tomkins. Over 525 signs developed by Sioux and other tribes. Written instructions and diagrams. Also 290 pictographs. 111pp. 6⅛ x 9¼. 22029-X Pa. $3.95

THE WIT AND HUMOR OF OSCAR WILDE, Alvin Redman (ed.). More than 1,000 ripostes, paradoxes, wisecracks: Work is the curse of the drinking classes; I can resist everything except temptation; etc. 258pp. 5⅜ x 8½. 20602-5 Pa. $6.95

SHAKESPEARE LEXICON AND QUOTATION DICTIONARY, Alexander Schmidt. Full definitions, locations, shades of meaning in every word in plays and poems. More than 50,000 exact quotations. 1,485pp. 6½ x 9¼. 2-vol. set.
Vol. 1: 22726-X Pa. $17.95
Vol. 2: 22727-8 Pa. $17.95

SELECTED POEMS, Emily Dickinson. Over 100 best-known, best-loved poems by one of America's foremost poets, reprinted from authoritative early editions. No comparable edition at this price. Index of first lines. 64pp. 5$\frac{3}{16}$ x 8¼.
26466-1 Pa. $1.00

THE INSIDIOUS DR. FU-MANCHU, Sax Rohmer. The first of the popular mystery series introduces a pair of English detectives to their archnemesis, the diabolical Dr. Fu-Manchu. Flavorful atmosphere, fast-paced action, and colorful characters enliven this classic of the genre. 208pp. 5$\frac{3}{16}$ x 8¼. 29898-1 Pa. $2.00

THE MALLEUS MALEFICARUM OF KRAMER AND SPRENGER, translated by Montague Summers. Full text of most important witchhunter's "bible," used by both Catholics and Protestants. 278pp. 6⅝ x 10. 22802-9 Pa. $12.95

SPANISH STORIES/CUENTOS ESPAÑOLES: A Dual-Language Book, Angel Flores (ed.). Unique format offers 13 great stories in Spanish by Cervantes, Borges, others. Faithful English translations on facing pages. 352pp. 5⅜ x 8½.
25399-6 Pa. $9.95

GARDEN CITY, LONG ISLAND, IN EARLY PHOTOGRAPHS, 1869–1919, Mildred H. Smith. Handsome treasury of 118 vintage pictures, accompanied by carefully researched captions, document the Garden City Hotel fire (1899), the Vanderbilt Cup Race (1908), the first airmail flight departing from the Nassau Boulevard Aerodrome (1911), and much more. 96pp. 8⅞ x 11¾. 40669-5 Pa. $12.95

OLD QUEENS, N.Y., IN EARLY PHOTOGRAPHS, Vincent F. Seyfried and William Asadorian. Over 160 rare photographs of Maspeth, Jamaica, Jackson Heights, and other areas. Vintage views of DeWitt Clinton mansion, 1939 World's Fair and more. Captions. 192pp. 8⅞ x 11. 26358-4 Pa. $14.95

CAPTURED BY THE INDIANS: 15 Firsthand Accounts, 1750-1870, Frederick Drimmer. Astounding true historical accounts of grisly torture, bloody conflicts, relentless pursuits, miraculous escapes and more, by people who lived to tell the tale. 384pp. 5⅜ x 8½. 24901-8 Pa. $9.95

THE WORLD'S GREAT SPEECHES (Fourth Enlarged Edition), Lewis Copeland, Lawrence W. Lamm, and Stephen J. McKenna. Nearly 300 speeches provide public speakers with a wealth of updated quotes and inspiration–from Pericles' funeral oration and William Jennings Bryan's "Cross of Gold Speech" to Malcolm X's powerful words on the Black Revolution and Earl of Spenser's tribute to his sister, Diana, Princess of Wales. 944pp. 5⅜ x 8⅜. 40903-1 Pa. $15.95

THE BOOK OF THE SWORD, Sir Richard F. Burton. Great Victorian scholar/adventurer's eloquent, erudite history of the "queen of weapons"–from prehistory to early Roman Empire. Evolution and development of early swords, variations (sabre, broadsword, cutlass, scimitar, etc.), much more. 336pp. 6⅛ x 9¼.
25434-8 Pa. $9.95

AUTOBIOGRAPHY: The Story of My Experiments with Truth, Mohandas K. Gandhi. Boyhood, legal studies, purification, the growth of the Satyagraha (nonviolent protest) movement. Critical, inspiring work of the man responsible for the freedom of India. 480pp. 5⅜ x 8½. (Available in U.S. only.) 24593-4 Pa. $9.95

CELTIC MYTHS AND LEGENDS, T. W. Rolleston. Masterful retelling of Irish and Welsh stories and tales. Cuchulain, King Arthur, Deirdre, the Grail, many more. First paperback edition. 58 full-page illustrations. 512pp. 5⅜ x 8½. 26507-2 Pa. $9.95

THE PRINCIPLES OF PSYCHOLOGY, William James. Famous long course complete, unabridged. Stream of thought, time perception, memory, experimental methods; great work decades ahead of its time. 94 figures. 1,391pp. 5⅜ x 8½. 2-vol. set.
Vol. I: 20381-6 Pa. $14.95
Vol. II: 20382-4 Pa. $16.95

THE WORLD AS WILL AND REPRESENTATION, Arthur Schopenhauer. Definitive English translation of Schopenhauer's life work, correcting more than 1,000 errors, omissions in earlier translations. Translated by E. F. J. Payne. Total of 1,269pp. 5⅜ x 8½. 2-vol. set. Vol. 1: 21761-2 Pa. $12.95
Vol. 2: 21762-0 Pa. $12.95

MAGIC AND MYSTERY IN TIBET, Madame Alexandra David-Neel. Experiences among lamas, magicians, sages, sorcerers, Bonpa wizards. A true psychic discovery. 32 illustrations. 321pp. 5⅜ x 8½. (Available in U.S. only.) 22682-4 Pa. $9.95

THE EGYPTIAN BOOK OF THE DEAD, E. A. Wallis Budge. Complete reproduction of Ani's papyrus, finest ever found. Full hieroglyphic text, interlinear transliteration, word-for-word translation, smooth translation. 533pp. 6½ x 9¼.
21866-X Pa. $12.95

MATHEMATICS FOR THE NONMATHEMATICIAN, Morris Kline. Detailed, college-level treatment of mathematics in cultural and historical context, with numerous exercises. Recommended Reading Lists. Tables. Numerous figures. 641pp. 5⅜ x 8½.
24823-2 Pa. $11.95

PROBABILISTIC METHODS IN THE THEORY OF STRUCTURES, Isaac Elishakoff. Well-written introduction covers the elements of the theory of probability from two or more random variables, the reliability of such multivariable structures, the theory of random function, Monte Carlo methods of treating problems incapable of exact solution, and more. Examples. 502pp. 5⅜ x 8½. 40691-1 Pa. $16.95

THE RIME OF THE ANCIENT MARINER, Gustave Doré, S. T. Coleridge. Doré's finest work; 34 plates capture moods, subtleties of poem. Flawless full-size reproductions printed on facing pages with authoritative text of poem. "Beautiful. Simply beautiful."–*Publisher's Weekly*. 77pp. 9¼ x 12. 22305-1 Pa. $7.95

NORTH AMERICAN INDIAN DESIGNS FOR ARTISTS AND CRAFTSPEOPLE, Eva Wilson. Over 360 authentic copyright-free designs adapted from Navajo blankets, Hopi pottery, Sioux buffalo hides, more. Geometrics, symbolic figures, plant and animal motifs, etc. 128pp. 8⅜ x 11. (Not for sale in the United Kingdom.) 25341-4 Pa. $9.95

SCULPTURE: Principles and Practice, Louis Slobodkin. Step-by-step approach to clay, plaster, metals, stone; classical and modern. 253 drawings, photos. 255pp. 8⅛ x 11.
22960-2 Pa. $11.95

THE INFLUENCE OF SEA POWER UPON HISTORY, 1660–1783, A. T. Mahan. Influential classic of naval history and tactics still used as text in war colleges. First paperback edition. 4 maps. 24 battle plans. 640pp. 5⅜ x 8½. 25509-3 Pa. $14.95

THE STORY OF THE TITANIC AS TOLD BY ITS SURVIVORS, Jack Winocour (ed.). What it was really like. Panic, despair, shocking inefficiency, and a little heroism. More thrilling than any fictional account. 26 illustrations. 320pp. 5⅜ x 8½. 20610-6 Pa. $8.95

FAIRY AND FOLK TALES OF THE IRISH PEASANTRY, William Butler Yeats (ed.). Treasury of 64 tales from the twilight world of Celtic myth and legend: "The Soul Cages," "The Kildare Pooka," "King O'Toole and his Goose," many more. Introduction and Notes by W. B. Yeats. 352pp. 5⅜ x 8½. 26941-8 Pa. $8.95

BUDDHIST MAHAYANA TEXTS, E. B. Cowell and others (eds.). Superb, accurate translations of basic documents in Mahayana Buddhism, highly important in history of religions. The Buddha-karita of Asvaghosha, Larger Sukhavativyuha, more. 448pp. 5⅜ x 8½. 25552-2 Pa. $12.95

ONE TWO THREE . . . INFINITY: Facts and Speculations of Science, George Gamow. Great physicist's fascinating, readable overview of contemporary science: number theory, relativity, fourth dimension, entropy, genes, atomic structure, much more. 128 illustrations. Index. 352pp. 5⅜ x 8½. 25664-2 Pa. $9.95

EXPERIMENTATION AND MEASUREMENT, W. J. Youden. Introductory manual explains laws of measurement in simple terms and offers tips for achieving accuracy and minimizing errors. Mathematics of measurement, use of instruments, experimenting with machines. 1994 edition. Foreword. Preface. Introduction. Epilogue. Selected Readings. Glossary. Index. Tables and figures. 128pp. 5³/₈ x 8¹/₂. 40451-X Pa. $6.95

DALÍ ON MODERN ART: The Cuckolds of Antiquated Modern Art, Salvador Dalí. Influential painter skewers modern art and its practitioners. Outrageous evaluations of Picasso, Cézanne, Turner, more. 15 renderings of paintings discussed. 44 calligraphic decorations by Dalí. 96pp. 5⅜ x 8½. (Available in U.S. only.) 29220-7 Pa. $5.95

ANTIQUE PLAYING CARDS: A Pictorial History, Henry René D'Allemagne. Over 900 elaborate, decorative images from rare playing cards (14th–20th centuries): Bacchus, death, dancing dogs, hunting scenes, royal coats of arms, players cheating, much more. 96pp. 9¼ x 12¼. 29265-7 Pa. $12.95

MAKING FURNITURE MASTERPIECES: 30 Projects with Measured Drawings, Franklin H. Gottshall. Step-by-step instructions, illustrations for constructing handsome, useful pieces, among them a Sheraton desk, Chippendale chair, Spanish desk, Queen Anne table and a William and Mary dressing mirror. 224pp. 8⅛ x 11¼. 29338-6 Pa. $16.95

THE FOSSIL BOOK: A Record of Prehistoric Life, Patricia V. Rich et al. Profusely illustrated definitive guide covers everything from single-celled organisms and dinosaurs to birds and mammals and the interplay between climate and man. Over 1,500 illustrations. 760pp. 7½ x 10⅛. 29371-8 Pa. $29.95

Prices subject to change without notice.

Available at your book dealer or write for free catalog to Dept. GI, Dover Publications, Inc., 31 East 2nd St., Mineola, N.Y. 11501. Dover publishes more than 500 books each year on science, elementary and advanced mathematics, biology, music, art, literary history, social sciences and other areas.